Lecture Notes in Mathematics

Edited by A. Dold and B. Eckmann

479

Stephen D. Fisher
Joseph W. Jerome

Minimum Norm Extremals in Function Spaces

With Applications to Classical and Modern Analysis

Springer-Verlag
Berlin · Heidelberg · New York 1975

Authors
Prof. Stephen D. Fisher
Department of Mathematics
Northwestern University
Evanston, Illinois 60201
U.S.A.

Prof. Joseph W. Jerome
Department of Mathematics
and the Technological Institute
Northwestern University
Evanston, Illinois 60201
U.S.A.

Library of Congress Cataloging in Publication Data

Fisher, Stephen D 1941-
Minimum norm extremals in function spaces with applications to classical and modern analysis.

(Lecture notes in mathematics ; 479)
Bibliography: p.
Includes index.
1. Calculus of variations. 2. Function spaces.
3. Approximation theory. I. Jerome, Joseph W., joint author. II. Title. III. Series: Lecture notes in mathematics (Berlin) ; 479.
QA3.L28 no. 479 [QA316] 510'.8s [515'.64] 75-23001

AMS Subject Classifications (1970): 34A10, 34B15, 35A15, 41A05, 41A10, 41A15, 41A25, 41A45, 41A65, 42A04, 46E15, 47B30, 47H15, 49A10, 49A25, 49A35, 49A40, 49A50, 49A99

ISBN 3-540-07394-9 Springer-Verlag Berlin · Heidelberg · New York
ISBN 0-387-07394-9 Springer-Verlag New York · Heidelberg · Berlin

Printed in Germany
Offsetdruck: Julius Beltz, Hemsbach/Bergstr.

For Naomi and Doreen with affection and appreciation.

PREFACE

This monograph has arisen out of the joint collaboration of the authors over the last few years. As such, the bulk of the material represents the authors' own research, obtained jointly or individually. The authors have exploited the techniques of functional analysis to provide applications in approximation theory, differential equations and control theory. In particular, the elastica and the bang-bang controls play a decisive role.

The authors wish to thank the National Science Foundation for their continued summer research support over this period. The second author is indebted to the British Science Research Council for sabbatical support during 1974-1975.

Finally, the authors express their admiration and affection for Garrett Birkhoff, Michael Golomb, George Lorentz and Iso Schoenberg, without whose prior efforts this monograph could not have been written.

Evanston, Illinois and
Oxford, England
December, 1974

Table of Contents

SECTION 0. INTRODUCTION

This monograph gives an account of recent developments in variational analysis and approximation theory which were largely unpublished prior to 1973. The material, however, has its roots deeply in the classical mathematical tradition and derives its point of view out of the solution of specifically formulated mathematical problems. Perhaps the dominant of these underlying problems is the determination of an elastica, defining a smooth curve in the plane subject to pin supports, which minimizes the strain energy quantity given by the integral of the squared curvature. A variant of this problem in which pressure is applied only at both ends of the elastica, was first considered by James Bernoulli in 1694 and completely solved by Euler in 1744 in his authoritative book which established the calculus of variations as a subject in its own right. Euler discovered that the configuration assumed by the elastica was expressible in terms of an elliptic integral. An understanding of the variational aspects of the general interpolation problem by elastica with minimum strain energy has been achieved only recently, however. This is due to the problem of existence; solutions do not exist, in general, as can be seen by smoothly piecing together arcs of circles near infinity to obtain an admissible interpolant with strain energy as small as desired. However, an a priori upper bound for the lengths of the admissible curves is a sufficient condition for the existence of an elastica with minimum strain energy, as has been conjectured by Garrett Birkhoff and established by the second author. This criterion of bounded length is thus a natural necessary and sufficient condition for existence of solutions and leads to an analysis via Lagrange multipliers, of the extremal solutions which has recently been achieved by the authors. An account of this may be found in Examples 1.1 and 1.2 of Chapter one and in Chapters eight and nine. The methods, primarily of functional analysis, which were used to achieve this understanding provide the perspective through which a range of minimum norm problems may be analyzed. Thus, as is frequent in mathematics, methods developed for special problems have more general application. Specifically, we are interested in this monograph in minimum norm problems of the form

$$\|Tf-g\| = \alpha = \inf\{\|Tu-g\| : u \in U\} \tag{0.1}$$

where T is in general a nonlinear mapping of a Banach space X into a Banach space Y, U is a subset of X and g is a fixed element in Y. It is required that $f \in U$. In Chapter one we develop an existence theory, the structure of which is comparable conceptually to the well-known theory of the minimization of a weakly lower semi-continuous functional over a weakly compact set. These theorems were originally developed by the authors to treat the problem of minimum curvature in L^p, $1 < p < \infty$ and in L^∞ for curves which are graphs of functions and preceded the work on the elastica. However, by the same theorems, are obtained in Examples 1.3, 1.4 and 1.5, existence theorems for nonlinear ordinary differential equations, nonlinear partial differential equations and optimal controls for systems governed by nonlinear equations. The results of Examples 1.3 and 1.4 are presented as Theorems 3.5 and 3.6 of Chapter three. The interest in these results lies as much in the method of proof as in the results themselves.

In Chapter two, we consider the special case when T is linear in (0.1). It is shown that all of the hypotheses of

Chapter one hold naturally in this setting. Our primary applications in this chapter are to spline-type solutions in one and several variables in the L^p-norms for $1 < p \leq \infty$. Chapters one and two form part one, the existence portion of the monograph.

Chapters three, four and five take up the question of characterization of solutions of (0.1). In the very general Chapter three, it is shown via duality methods that the element, for which $Tf - g$ achieves its norm α as a linear functional, is orthogonal to the image, under the Frechet derivative of T at f, of a linear subspace U_0 associated with U. This result gives in particular a rigorous foundation to the derivation of the Euler equation satisfied by f. Applications include the existence theorems on nonlinear differential equations cited earlier. A result is also presented in Theorem 3.7 which gives a complete characterization for solutions of (0.1) when T is a linear differential operator, $Y = L^p$, $1 < p < \infty$, X is the corresponding Sobolev space and U is defined by linear inequality constraints involving linear combinations of derivatives at nodal points. Additional applications of Chapter three appear in Chapters six, eight and ten. In Chapter four, T is taken to be an elliptic operator of even order and minimization is in the space $L^\infty(\Omega)$. Here the first instance of bang-bang phenomena appears in the monograph; f is unique and $|Tf-g|$ is of constant modulus on Ω. In Chapter five, a comprehensive analysis of minimization in L^1 is undertaken. Special techniques are required since L^1 is not a dual space and it is found that embedding L^1 in the space NBV leads to solutions which are integral transforms of point masses. These results hold in one and several variables.

Chapter six studies the question of uniqueness in L^∞ minimization problems. Although uniqueness does not hold in general in a global sense, there are local theorems expressed in terms of core subsets of uniqueness in Euclidean space. The first three parts of the monograph, concluding with Chapter six, give an account primarily of the authors' own work.

Part four, consisting of Chapter seven, is an analysis of bang-bang optimal controls in the L^∞ norm for systems governed by linear ordinary differential equations and satisfying multipoint inequality constraints. Here we have been guided by the work of Donald McClure and have developed his presentation. The theorems giving rise to bang-bang controls, with a finite number of discontinuities, are extremely general and require minimal hypotheses on the linear systems and virtually no hypotheses on the inequality constraints. Theorem 7.5 is perhaps the fundamental theorem of the chapter. We do not assume that the linear system is equivalent to an nth order differential equation.

As remarked earlier, Chapters eight and nine are results of the authors which are concerned primarily with Lagrange multipliers and the problem of minimum curvature. They are independent and Chapter nine is by far the deeper. As an auxiliary result, we obtain the differential equation, in arc length, which is satisfied locally by the elastica: $\ddot{\kappa} + \frac{\kappa^3}{2} - \frac{\lambda}{2}\kappa = 0$, where λ is a nonnegative number. In Chapter nine we also examine minimum curvature in L^∞ and obtain the result that a solution exists which is obtained by piecing together arcs of circles. Part six, consisting of Chapter ten, analyzes the convergence to smooth functions of solutions of (0.1) with $g = 0$ and T a nonlinear differential operator. The important result obtained here (Theorem 10.3) is that convergence orders for nonlinear T are the same as for linear T. In this chapter, we also consider the general problem of the minimization of a (symmetric) quadratic form which need not be nonnegative. The existence theory here makes use of the Riesz-Fredholm-Schauder theory and holds very generally for projections with respect to non-symmetric bilinear forms. The results apply equally well in one or several Euclidean dimensions. A convergence result, which makes use of the discrete Fourier transform, is presented in Theorem 10.13. A final section of this chapter discusses convergence for the NBV extremals of Chapter five. This chapter also represents the authors' own work.

Chapter eleven commences part seven of the monograph, devoted to an exposition of a rich variety of L^∞ extremal problems in which perfect spline functions arise as extremals. This role is so striking that we shall summarize briefly the remarkable circle of ideas. In 1937, there appeared independently two published proofs of a result now known as the Favard-Achieser-Krein theorem on the best approximation of smooth 2π-periodic functions by trigonometric polynomials of degree m. It was found that, among such n-fold integrals of functions in the unit ball of $L^\infty(0,2\pi)$, the functions at a maximum distance from the trigonometric polynomials of degree m are (essentially) translates of a perfect spline function of degree n. Perfect here means that the nth derivative is of constant modulus, in this case of modulus one. Specifically, the extremals are 2π-periodic functions of the form

$$f_{nm}(x) = \left(\frac{1}{m+1}\right)^n f_n((m+1)x)$$

where f_n is the perfect spline function of mean value zero for which

$$f_n^{(n)}(x) = \begin{cases} \text{signum } \cos x, & n \text{ even}, \\ \text{signum } \sin x, & n \text{ odd}. \end{cases}$$

In 1939, Kolmogorov published a complete solution of the Landau problem on the real line. Specifically, he calculated the smallest numbers $C_{n,\nu}$ for which

$$(0.2) \qquad \|F^{(\nu)}\| \leq C_{n,\nu}\|F\|^{1-\nu/n}\|F^{(n)}\|^{\nu/n}, \quad 0 < \nu < n,$$

where the norm is the L^{∞} norm taken on $\mathbb{R}$. Kolmogorov showed that the extremals for (0.2) are (essentially) the perfect spline functions

$$E_n(x) = f_n(\pi x)/K_n$$

where f_n is given above and $K_n = \|f_n\|$; i.e., equality holds in (0.2) for $F = E_n$. Also,

$$C_{n,\nu} = K_{n-\nu}/K_n^{1-\nu/n}, \quad 0 < \nu < n, \quad n = 2,3,\ldots,$$

and

$$K_n = \frac{4}{\pi}\sum_{k=0}^{\infty}\frac{(-1)^{k(n+1)}}{(2k+1)^{n+1}}, \quad n = 1,2,\ldots .$$

In Chapter eleven we present a new proof, by duality, of the Favard-Achieser-Krein result. We also consider the corresponding problem in the context of approximation on a compact interval by algebraic polynomials. A similar proof by duality, recently obtained by the first author, establishes that perfect spline functions are extremals in this problem as well; in general, these will not have the uniform knot spacing of the periodic extremals, although the exact location of these knots remains an open problem. In Chapter eleven we also present a new proof of Bernstein's result on the asymptotic behavior of the best approximation constants in the algebraic problem; this result is then used to deduce a generalization of the Favard-Achieser-Krein theorem, due to Krein, in which functions of exponential type are used as approximants on the real line.

The theorem of Kolmogorov, as proved by A. Cavaretta, is presented in Chapter thirteen. Chapter twelve is a transition chapter in which the properties of the functions E_n above, required in Chapter thirteen, are derived in detail. Schoenberg has called these functions Euler splines since they are essentially periodic extensions of the Euler polynomials. In Chapter twelve we also give still a third extremal problem for which the Euler splines are extremals, viz, the interpolation of values $(-1)^i$ at the integers i, subject to minimal nth derivative norm. This result is due to the authors. We also cite here results of Schoenberg on the unique interpolation of bounded data on $\mathbb{R}$ by bounded spline functions. Chapter fourteen concludes this part of the monograph by presenting the theorem of Karlin, as proved by De Boor, on the interpolation of n + r data values, with minimal nth derivative norm at n + r points of multiplicity at most n, by a perfect spline function of degree n with fewer than r (simple) knots. Although the generality of Chapter seven is far greater, the merit of this result lies in the global bound of r - 1 for the number of knots.

The final part contains the summary Chapter fifteen of several different topics. For example, a different solution to the problem of Chapter fourteen, due to Favard, is described. This solution has the merit that it is the limit of the (unique) L^p extremal solutions as $p \to \infty$. A result of Tihomirov, on the n-widths of smooth function classes, is described. Here perfect spline knot locations define best approximating subspaces.

Finally, some results of Golomb on the possibility of extending functions smoothly off arbitrary sets will be presented. These results complement the Whitney results. There follows a chapter giving an account of the generalized spline theory and the monograph closes with the epilogue Chapter seventeen.

The documentation of the mongraph assumes the form of sectional bibliographies, but does not preclude items from being referenced in subsequent chapters. Any nonstandard facts in functional analysis have been documented. We have assumed familiarity, however, with such facts as the equivalence of a reflexive Banach space with the property of possessing a sequentially weakly compact unit ball; of the property that the dual of a separable normed linear space has a sequentially weak-* compact unit ball; and of the fact that a closed, convex subset of a Banach space is weakly closed. Such facts and others appear in the various books on functional analysis which we have cited. We also assume familiarity with the various duality properties satisfied by the Lebesgue spaces $L^p(\Omega,\mu)$, $1 \leq p \leq \infty$ where (Ω,μ) is a σ-finite measure space. In fact, unless further restricted, (Ω,μ) will always refer to such. However, most of our applications to the Lebesgue spaces involve domains Ω in Euclidean space. We also assume familiarity with basic properties of distributions.

We shall define some elementary terms used frequently in the sequel. A bang-bang control function is one of constant modulus. A bang-bang or perfect spline function of degree n is a piecewise polynomial (of degree n) in C^{n-1} possessing an nth (discontinuous) derivative of constant modulus. More generally, a bang-bang or perfect spline function s satisfies $|Ls|$ is constant, where L is a linear differential operator in one or several variables whose fundamental solutions are employed locally to construct s. By a flat U we shall mean a translate $U_0 + u_0$ of a linear subspace U_0. U_0 is said to be determined by a collection of linear functionals if U_0 is the intersection of their kernels. An extended Hermite-Birkhoff (EHB) linear functional is one whose application to a smooth function is calculated by some prescribed linear combination of derivatives, through a fixed order, at a specified point.

The Sobolev spaces in one and several variables play a fundamental role in the sequel and we shall define them for completeness. For $n \geq 0$, $1 \leq p \leq \infty$ and Ω a domain in $\mathbb{R}^N$ the (usually real) Sobolev spaces of integral order are defined:

$$W^{n,p}(\Omega) = \{f : D^\gamma f \in L^p(\Omega) : 0 \leq |\gamma| \leq n\}.$$

Here $\gamma = (\gamma_1, \ldots, \gamma_N)$ is a standard mult-integer with nonnegative entries and $D^\gamma = \left(\frac{\partial}{\partial x_1}\right)^{\gamma_1} \cdots \left(\frac{\partial}{\partial x_N}\right)^{\gamma_N}$ is taken in the sense of distributions; $|\gamma| = \sum_i \gamma_i$. Thus, the elements of $W^{n,p}(\Omega)$ have distribution derivatives through order n in $L^p(\Omega)$. A particularly simple description is available when $N = 1$: $W^{n,p}(\Omega)$ consists of n-fold indefinite integrals of $L^p(\Omega)$ functions. Norms and semi-norms are defined in $W^{n,p}(\Omega)$ as follows. Single braces refer to semi-norms and double braces to norms.

$$|f|_{\ell,p} = \begin{cases} \{\int_\Omega \sum_{|\alpha| = \ell} |D^\alpha f|^p\}^{1/p}, & 1 \leq p < \infty, \\ \\ \max_{|\alpha| = \ell} (\operatorname{ess\,sup}_{x \in \Omega} |D^\alpha f(x)|), & p = \infty, \end{cases}$$

$$\|f\|_{n,p} = \begin{cases} \sum_{\ell=0}^{n} \{|f|_{\ell,p}^{p}\}^{1/p}, & p < \infty, \\ \max_{0 \le \ell \le n} |f|_{\ell,\infty}, & p = \infty. \end{cases}$$

$W_0^{n,p}(\Omega)$ is the completion in $W^{n,p}(\Omega)$ of the class $C_0^\infty(\Omega)$ of infinitely differentiable functions with compact support in Ω. We shall assume as known the facts that $W^{n,p}(\Omega)$, $1 \le p \le \infty$, are Banach spaces, reflexive if $1 < p < \infty$. A standard fact about $W_0^{n,p}(\Omega)$ is that the semi-norm of order n defines a norm equivalent to that induced by the norm in $W^{n,p}(\Omega)$. We shall make use of this when Ω is an interval of the real line.

We shall be as economical as possible in the use of notation. Thus, we shall frequently omit all norm subscripts when the contextual meaning is clear; for functions of one variable a single norm subscript indicates which L^p norm, $1 \le p \le \infty$, is being utilized. For functions of several variables, we shall abbreviate $W^{n,2}(\Omega)$ to $W^n(\Omega)$ with a corresponding norm subscript omission. The reader should experience no difficulty with these omissions and should benefit from the resultant notational simplicity.

We close the introduction with a remark concerning the Euler splines E_n, referred to earlier. These seem destined to become, ultimately, part of the standard collection of special functions if not already. At the least, the Euler spline seems to play the role in the twentieth century of the cycloid in the seventeenth century. This latter function was found in 1697 to be the solution curve of the brachistochrone problem which followed an earlier observation by Huygens (1673) that a pendulum bob, swinging along a cycloidal arc, takes the same time to complete swings of large and of small amplitude.

REFERENCE

Morris Kline, Mathematical Thought from Ancient to Modern Times, Oxford University Press, Oxford, England, 1972.

PART I. EXISTENCE THEOREMS

§1. Nonlinear Minimization Problems

1.1 General Theorems

In this section, we isolate the general hypotheses required for the existence of a solution of the minimization problem

$$(1.1) \qquad \|Tu_0-y\| = \alpha = \inf_{u\in U}\|Tu-y\|$$

where T is an operator mapping a normed linear space X into a normed linear space Y and U is a subset of X.

Definition 1.1 An operator T mapping a normed linear space X into a normed linear space Y is said to have property (D) if, whenever $\{x_n\}$ is a bounded sequence in X for which $\{Tx_n\}$ is weakly convergent in Y, then some subsequence of $\{x_n\}$ is weakly convergent in X.

A weakly continuous operator satisfying property (D) maps weakly closed, bounded subsets of X onto weakly closed subsets of Y; in this case, I - T is (weakly) demicompact in the terminology of [1.10].

Theorem 1.1. Let X and Y be separable normed linear spaces with Y reflexive and let T be an operator mapping X into Y with property (D) and suppose also that $\{Tx_n\}$ is weakly convergent to Tx in Y whenever $\{x_n\}$ is weakly convergent to x in X. Let U be a weakly closed subset of X, let $y \in Y$, and consider the minimization problem (1.1). This problem has a solution if there is a bounded sequence $\{x_n\}$ in U with $\alpha = \inf\|Tx_n-y\| : n = 1,2,\ldots\}$, i.e., a bounded minimizing sequence in U.

Proof: The hypotheses have been explicitly designed to yield an immediate conclusion. Thus, if $\{x_n\}$ is a bounded minimizing sequence, then, in particular, $\{Tx_n\}$ is bounded. Since the unit ball of Y is weakly sequentially compact some subsequence, again denoted by $\{Tx_n\}$, converges to an element $z \in Y$. Since T satisfies property (D) there is a subsequence of $\{x_n\}$, again denoted by $\{x_n\}$, such that $\{x_n\}$ is weakly convergent to an element $x \in U$. Since T is weakly sequentially continuous, $\{Tx_n\}$ converges weakly to Tx. Hence, $z = Tx$ and, by the lower semicontinuity of the norm with respect to weak convergence,

$$\|Tx-y\| \leq \liminf_{n \to \infty} \|Tx_n - y\| = \alpha .$$

Hence, $\|Tx-y\| = \alpha$ as desired. □

Thus, to ensure the existence of a solution of (1.1) we must find conditions which guarantee that

(1.2) the operator T has property (D);

(1.3) T is weakly continuous;

(1.4) there exists a bounded minimizing sequence in U.

Note that, if X is a reflexive normed linear space, then (1.2) holds. Also, if U is bounded, then (1.4) holds a fortiori. Finally, there is a result analogous to Theorem 1.1 if the weak-* topology is utilized in X and Y. We state this as

Theorem 1.2. Let X and Y be Banach spaces which are the dual spaces of separable normed linear spaces. Let T be an operator mapping X into Y such that $\{Tx_n\}$ is weak-* convergent in Y to Tx whenever $\{x_n\}$ is weak-* convergent to x in X. Let U be a weak-* closed subset of X, let $y \in Y$ and consider the minimization problem (1.1). This problem has a solution if there is a bounded minimizing sequence in U.

Proof: Note that the unit balls of X and Y are weak-* sequentially compact. In particular, property (D) holds with respect to the weak-* topologies in X and Y. Note also the lower semicontinuity of the norm in Y with respect to weak-* convergence. The conclusion follows as in Theorem 1.1. □

It is frequently desirable to draw the conclusion of Theorem 1.2 without reference to the topology of X. The following corollary is immediate from the proof of Theorem 1.2.

Corollary 1.3. Let X be a linear space and let Y be a Banach space which is the dual of a separable normed linear space W. Let T be an operator mapping X into Y and let U be a subset of X. Consider the minimization problem (1.1). This problem has a solution if the intersection of TU with every closed ball centered at y is weak-* sequentially closed.

Remarks. These are standard results in nonlinear functional analysis. Note that condition (1.3) implies that the composition functional $\|\cdot\|\circ T$ is lower semicontinuous on U and that (1.2) and (1.4) are a substitute for the weak compactness of U. Viewed in this way, the general minimization problems of this section are simply variants of the standard minimization of a weakly lower semicontinuous functional over a weakly compact set in a normed linear space. This subject has been studied in depth by many authors, e.g., Daniel [1.3], Mikhlin [1.9] and Vainberg [1.11]. In the setting of control theory, which we touch upon briefly in this monograph, cf Berkovitz [1.2] and Lions [1.8].

We mention finally that a somewhat stronger version of Theorem 1.1 holds, in which it is required only that T map some subsequence of every weakly convergent sequence onto a weakly convergent sequence.

1.2 Applications

Example 1.1. The Problem of Minimum Curvature in L^2

One of the fundamental applications of the theory developed in section 1.1 is the verification of the existence of a smooth interpolating curve in the plane, passing through a prescribed set of points, with prescribed length L and minimum curvature or strain energy.

Thus, let X be the direct sum of two copies of $W^{2,2}(a,b)$, let Y be the direct sum of two copies of $L^2(a,b)$ and define T by

$$T(x,y) = (x'',y'').$$

Let points $\mathscr{P} = \{(x^i,y^i)\}_{i=0}^N$ be prescribed in the plane and define U to consist of all pairs $(x,y) \in X$ such that the planar curve $t \to (x(t),y(t))$, $a \leq t \leq b$, passes through the set $\mathscr{P}$, and satisfies the relation $(x')^2 + (y')^2 \equiv 1$. The problem of minimum curvature corresponds precisely to the case $a = 0$, $b = L$ and t the arc length parameter. Now X and Y are reflexive so that (1.2) holds and the unit ball of Y is weakly sequentially compact. To verify (1.3) note that T is continuous and linear and hence weakly continuous. To verify (1.4), we note that it only remains to show that, for every minimizing sequence (x_n,y_n) in X, $\{x_n\}$ and $\{y_n\}$ are uniformly bounded sequences (in the supremum norm). Choosing a fixed point in $\mathscr{P}$, say (x^0,y^0), we select $\{t_n\} \subset [a,b]$ so that $(x_n(t_n),y_n(t_n)) = (x^0,y^0)$, $n = 1, 2, \ldots$. Thus,

$$x_n(t) = \int_{t_n}^{t} x_n'(\tau)\,d\tau + x^0, \quad y_n(t) = \int_{t_n}^{t} y_n'(\tau)\,d\tau + y^0,$$

so that

$$|x_n(t)| \le (b-a) + |x^0|, \quad |y_n(t)| \le (b-a) + |y^0|, \quad a \le t \le b,$$

and (1.4) holds. It remains to show that U is weakly closed. Suppose that $(x_n, y_n) \to (x,y)$ weakly in X with $\{(x_n, y_n)\} \subset U$. By the uniform convergence of x_n' and y_n' to x' and y' respectively, it follows that $(x')^2 + (y')^2 \equiv 1$. We must show that the curve $(x(t), y(t))$, $a \le t \le b$, passes through the points of $\mathscr{P}$. Let $(x_*, y_*) \in \mathscr{P}$ and select $\{t_n\} \subset [a,b]$ so that $(x_n(t_n), y_n(t_n)) = (x_*, y_*)$. Let t_* be an accumulation point of $\{t_n\}$ in $[a,b]$. Then $(x(t_*), y(t_*)) = (x_*, y_*)$. Indeed, if $\varepsilon > 0$, then by the equicontinuity and uniform convergence of $\{x_n\}$ we have, for some n,

$$\begin{aligned} |x(t_*) - x_*| &\le |x(t_*) - x_n(t_*)| + |x_n(t_*) - x_n(t_n)| \\ &\le \varepsilon/2 + \varepsilon/2 \\ &\le \varepsilon \end{aligned}$$

with a similar statement for y. U is weakly closed, therefore, and we conclude by Theorem 1.1 that there is a pair $(x,y) \in U$ such that

$$(1.6) \qquad \int_a^b [x''^2(t) + y''^2(t)]dt = \inf_{(u,v)\in U} \int_a^b [u''^2(t) + v''^2(t)]dt.$$

We shall derive the properties of any such solution (x,y) in §9.

Example 1.2. Minimum Curvature in L^∞

Take X to be the direct sum of two copies of $W^{2,\infty}(a,b)$ and Y to be $L^\infty(a,b)$ and T to be

$$T(x,y) = (x'')^2 + (y'')^2$$

Define $U \subset X$ to consist of all pairs (x,y) such that the planar curve

$t \to (x(t),y(t))$, $a \leq t \leq b$, passes through a given finite point set $\mathscr{P}$ and such that $x'^2 + y'^2 \equiv 1$. The verification of the existence of a pair $(x,y) \in X$ satisfying

$$(1.7) \qquad \|(x'')^2+(y'')^2\|_\infty = \inf_{(u,v)\in U} \|(u'')^2+(v'')^2\|_\infty$$

follows readily from Corollary 1.3. Indeed, it clearly suffices to show that $(x_n'')^2 + (y_n'')^2 \to z$(weak-$*$) implies $z \in TU$ if $\{(x_n,y_n)\} \subset U$. However, since $\{x_n''\}$, $\{y_n''\}$, $\{x_n''^2\}$ and $\{y_n''^2\}$ are bounded sequences in Y, there exists a sequence of positive integers n_k for which $\{x_{n_k}''\}$, $\{y_{n_k}''\}$, $\{x_{n_k}''^2\}$, $\{y_{n_k}''^2\}$ are weak-$*$ convergent in Y to x_*, y_*, x_*^2, y_*^2. The verification that $(x_*,y_*) \in U$ is a consequence of the uniform convergence and equicontinuity of the sequences $\{x_n\}$ and $\{y_n\}$, and the uniform convergence of $\{x_n'\}$ and $\{y_n'\}$. Since the details were presented in the previous example, we omit them.

We have thus shown the existence of a smooth interpolating curve in the plane, passing through a prescribed set of points, with prescribed length L and minimum pointwise curvature. Indeed, this is the case $a = 0$, $b = L$ with t the arc length parameter. We shall derive the properties of any such solution in §9.

Example 1.3. Nonlinear Ordinary Differential Equations

Let φ and Ψ be real-valued C^1 functions on $[a,b] \times \mathbb{R}^N$. Let T be defined on $X = W^{n,p}(a,b)$, $1 < p < \infty$, by

$$Tf(t) = \varphi(t,f(t),\ldots,f^{(n-1)}(t))f^{(n)}(t) + \Psi(t,f(t),\ldots,f^{(n-1)}(t))$$

with $Y = L^p(a,b)$. Let U be a flat in X of codimension n, specified by (generalized) interpolation. If we make the assumptions that $\varphi \geq \delta > 0$ on $[a,b] \times \mathbb{R}^n$ and that Ψ is bounded on $\mathbb{R}^n$, then it is elementary to see that a bounded minimizing sequence for (1.1) exists, provided that U_0, the subspace of codimension n parallel to U, contains no polynomial of degree $\leq n - 1$ except the trivial one. T is weakly continuous since D^n is weakly continuous, and since weak convergence in X implies uniform convergence of derivatives through order n - 1 on [a,b]. Since U is closed and convex, U is weakly closed and hence the hypotheses of Theorem 1.1 are satisfied (note that X and Y are reflexive). As we shall see in §3 if the linearized problem has a solution, the constant α of (1.1) is 0 under these aforementioned hypotheses and hence the differential equation

$$Tf = g \qquad f \in U \tag{1.8}$$

has a solution $f \in W^{n,p}(a,b)$ for any $g \in L^p(a,b)$. If $p = \infty$ and U is weak-* closed, a similar result holds. Also, less restrictive hypotheses will be shown to insure a solution.

Example 1.4. Nonlinear Partial Differential Equations

Let Ω be a bounded, open set in $\mathbb{R}^m$ for which the Sobolev embedding theorem holds [1.1, p. 32], let L be a continuous, bijective mapping

$$L : W^{2k,2}(\Omega) \cap W_0^{k,2}(\Omega) \to L^2(\Omega)$$

where it is assumed that $2k > [m/2]$ and let Ψ be a $C^1(\overline{\Omega} \times \mathbb{R}^\ell)$ function, where ℓ is an integer equal to the dimension of the space of polynomials of degree $\leq \nu - 1$, with $\nu = 2k - [m/2]$. Let T be defined on $X = W^{2k,2}(\Omega) \cap W_0^{k,2}(\Omega)$ by

$$Tf(t) = Lf(t) + \Psi(t, D^\gamma f(t)),\ t \in \Omega,$$

where γ ranges over all multi-indices $\gamma = (\gamma_1,\ldots,\gamma_m)$ of length $|\gamma| \leq \nu - 1$. Let U = X and notice that if Ψ is a bounded function, then every minimizing sequence for (1.1) is bounded in X. Now L is a continuous linear mapping and hence weakly continuous and the mapping defined by Ψ is weakly continuous since weak convergence by the Sobolev embedding theorem, implies uniform convergence of derivatives through order $\nu - 1$. Thus (1.1) has a solution under the hypotheses. We shall show in §3 that this implies that the differential equation

$$\text{(1.9)} \qquad \begin{aligned} Tf &= g \\ f &\in X \end{aligned}$$

has a solution f, provided the linearized problem always has a solution.

<u>Example 1.5</u>. Optimal Control

Let $X = W_N^{1,p}(a,b)$ and $Y = L^p(a,b)$, $1 < p < \infty$, and let V be a weakly closed control set in Y. Let J be a functional defined on X × Y by

$$\text{(1.10)} \qquad J(x,u) = \int_a^b |f(t,x_1(t),\ldots,x_N(t),u(t))|^p dt,$$

where f is a C^1 function on $[a,b] \times \mathbb{R}^{N+1}$. A standard problem in the theory of optimal control of systems governed by ordinary differential equations is the minimization of (1.10) subject to

$$\text{(1.11)} \qquad \begin{aligned} x' &= Ax + Bu \\ x(a) &= x_0 \\ u &\in V \end{aligned}$$

where A is an N × N matrix of C[a,b] functions and B an N × 1 matrix of C[a,b] functions of which at least one is strictly of one sign.

This problem can easily be cast into the framework of section 1.1 as follows. If the N x N matrix Ψ is a (fundamental) solution of

$$\Psi' = A\Psi$$
$$\Psi(a) = I$$

then there is a one to one correspondence between $u \in V$ and $x \in X$ satisfying (1.11) which is given by the relation,

$$x(t) = \Psi(t)x_0 + \Psi(t) \int_a^t \Psi^{-1}(s)Bu(s)ds.$$

Clearly, the preimage of V under (1.11) is weakly closed in X and we denote this preimage by U. We may define a mapping $T : U \to Y$ by

$$Tx(t) = f(t, x_1(t), \ldots, x_N(t), u(t)).$$

If V is bounded, then U is bounded and it follows that (1.2), (1.3), and (1.4) are satisfied. Thus, the functional of (1.10) attains a minimum for some control $u \in V$ by Theorem 1.1.

Example 1.6. A Logarithm Solution of an L^∞ Problem.

Let $X = W^{1,\infty}(0,1)$, $Y = L^\infty(0,1)$, $Tf = f'e^f$ and let $U = \{f \in X : f(0) = 0, f(1) = 1\}$. Now note the representation,

$$f(t) = \ell n[1 + \int_0^t Tf(\tau)d\tau], \ 0 \leq t \leq 1$$

for each $f \in U$. If $f_n' \to f'$ in the weak-* topology of L^∞ and $f_n \in U$, then clearly, $f_n \to f$ uniformly and hence $f \in U$ and T is weak-* continuous. We shall show that every minimizing sequence for (1.1) is bounded. Indeed, since $f(x) = x$, $0 \leq x \leq 1$, is in U, it follows that $\alpha \leq e$. Suppose that $\|Tf\| \leq e + e^{-1}$. We shall show that f is uniformly bounded by 3 and hence $\|f\| \leq 2e^4$, i.e., every minimizing sequence is bounded. If $0 \leq t < e^{-2}$ then

$$e^{f(t)} - 1 = \int_0^t Tf(\tau)\,d\tau$$

and thus $|e^{f(t)}-1| \leq 2et < 2e^{-1}$ so that

$$\ell n(1-2e^{-1}) \leq f(t) \leq \ell n(1+2e^{-1}).$$

If $e^{-2} \leq t \leq 1$, then

$$e - e^{f(t)} = \int_t^1 Tf(\tau)\,d\tau$$

so that $|e-e^{f(t)}| \leq (e+e^{-1})(1-e^{-2})$. It follows that $|f(t)| \leq 3$ for $0 \leq t \leq 1$ and a solution of (1.1) exists by Theorem 1.2.

In §3 we shall show that f is unique and give the explicit representation

$$f(t) = \ell n[(e-1)t+1], \quad 0 \leq t \leq 1.$$

Remarks. Much of the material of this chapter is taken from the authors' papers [1.4, 1.5, 1.6, 1.7]. Example 1.1 is elaborated in detail in [1.7] where it is pointed out that an upper bound for the lengths of the admissible interpolating curves is a necessary and sufficient condition for the existence of a smooth interpolating curve of minimum curvature. Example 1.3 was introduced in [1.4], whereas Example 1.6 was discussed in [1.6] and an explicit solution given in [1.4]. The existence of a minimum for Example 1.3, independent of its application to differential equations, was presented in [1.6].

REFERENCES

1.1 S. Agmon, Lectures on Elliptic Boundary Value Problems, Van Nostrand, Princeton, 1965.

1.2 L. D. Berkovitz, Optimal Control Theory, Applied Mathematical Sciences 12, Springer-Verlag, 1974.

1.3 J. W. Daniel, The Approximate Minimization of Functionals, Prentice-Hall, Englewood Cliffs, N. J., 1971.

1.4 S. D. Fisher, "Solutions of some non-linear variational problems in L^{∞} and the problem of minimum curvature, Archives of Rational Mechanics and Analysis," to appear.

1.5 J. W. Jerome, "Minimization problems and linear and non-linear spline functions I: Existence," SIAM J. Numer. Anal., 10 (1973), 808-819.

1.6 __________, "Minimization problems and linear and non-linear spline functions, II: Convergence," SIAM J. Numer. Anal., 10 (1973), 820-830.

1.7 __________, "Smooth interpolating curves of prescribed length and minimum curvature," Proc. Amer. Math. Soc., 50 (1975).

1.8 J. L. Lions, Optimal Control of Systems Governed by Partial Differential Equations, Springer-Verlag, New York, 1971.

1.9 S. Mikhlin, The Minimization of a Quadratic Functional, Holden-Day, San Francisco, 1964.

1.10 W. V. Petryshyn, "Construction of fixed points of demicompact mappings in Hilbert space," J. Math. Anal. Appl., 14 (1966), 274-284.

1.11 M. M. Vainberg, Variational Methods for the Study of Nonlinear Operators, Holden-Day, San Francisco, 1964.

§2. Minimization with Linear Operators

2.1 General Theorems

In this chapter we specialize to the case where T is a linear operator in the minimization problem (1.1). We note the fact [2.4, p. 422] that for a linear operator T mapping one Banach space into another, continuity coincides with weak continuity.

Theorem 2.1. Let X and Y be Banach spaces and let T be a continuous linear mapping of X into Y with closed range. If the null space of T is finite dimensional, then T has property (D).

Proof: We may assume that T maps X onto Y. Let N be the null space of T. N is complemented in X; there is a closed subspace M of X with $X = N \oplus M$. Let S be the restriction of T to M, so that S maps M both 1 - 1 and onto Y and so has a continuous inverse. Let $\{x_k\}$ be a bounded sequence in X with $\{Tx_k\}$ weakly convergent to Tx_0 in Y. If $x_k = n_k + m_k$, then $Tx_k = Tm_k = Sm_k$ so that $\{Sm_k\}$ is weakly convergent to $Tx_0 = Sm_0$. Hence, $\{m_k\}$ is weak convergent to m_0 in M. Since $\{n_k\}$ is a bounded sequence in N which is finite dimensional we may assume that $n_k \to n$ in N. Hence, $x_k \to n + m_0$ weakly in X and the proof is complete. □

Theorem 2.2. Let X and Y be Banach spaces and suppose that Y is reflexive. Let T be a linear continuous mapping of X onto Y satisfying property (D). Let Γ be a continuous linear mapping of X onto a Banach space Z and let K be a closed, bounded subset of Z. Define

$$U = \{x \in X : \Gamma x \in K\}.$$

Then, for each $y \in Y$, the minimization problem

$$\alpha = \inf\{\|Tx-y\| : x \in U\} \tag{2.1}$$

has a solution. In particular, (2.1) has a solution if T is a continuous bijection and either X is reflexive or T has finite dimensional null space.

Proof: By Theorem 1.1, it suffices to show that there is a bounded minimizing sequence in U for (2.1). If $\{x_k\}$ is a minimizing sequence, we may assume that $\{Tx_k\}$ is weakly convergent in Y by choosing an appropriate subsequence. Thus, we may conclude that $\{[x_k]\}$ is bounded in X/N and hence $\{x_k+y_k\}$ is bounded in X for a sequence $\{y_k\}$ in N. Since both $\{\Gamma(x_k+y_k)\}$ and $\{\Gamma x_k\}$ are bounded sequences in Z, so is $\{\Gamma y_k\}$. Using the quotient space N/M, where $M = \{y \in N : \Gamma y = 0\}$, it follows we can obtain elements y_k', $\Gamma y_k' = \Gamma y_k$, such that $\|y_k'\|$ is bounded. It follows that $\{x_k+y_k-y_k'\}$ is bounded and is contained in U. Since $T(x_k+y_k-y_k') = Tx_k$, it follows that we have produced a bounded minimizing sequence. □

Let X and Y be Banach spaces and let T be a continuous linear mapping of X onto Y. Let U_0 be a closed linear subspace of X. Consider the following statements.

(2.2)

(i) The null space of T is finite dimensional

(ii) T satisfies property (D) with respect to strong convergence in X and Y.

(iii) TU_0 is closed.

(iv) The minimization problem (2.1) has a solution.

Theorem 2.3. The implications (i) ⟹ (ii) ⟹ (iii) hold. If Y is reflexive, then (iii) ⟹ (iv).

Proof: (i) ⟹ (ii) Let $\{x_n\}$ be bounded in X and $\{Tx_n\}$ convergent in Y. The mapping $T_0 : X/N \to Y$ defined by

$$T_0[x] = Tx$$

is linear, continuous and bijective and hence has a continuous inverse. Thus, there exists a sequence $\{t_n\}$ in N such that $\{x_n+t_n\}$ is convergent in X. Since $\{t_n\}$ is bounded and N is locally compact, there is a subsequence $\{t_{n_k}\}$ convergent in X. It follows that $\{x_{n_k}\}$ is convergent in X and (ii) holds.

(ii) $\Longrightarrow$ (iii) It is clear that, if (ii) holds, T maps bounded, closed sets onto closed sets. If TU_0 is not closed, then there is a sequence $\{[x_n]\}$ in $U_0/U_0 \cap N$ such that

$$(2.3) \qquad \|[x_n]\| = 1, \; Tx_n \to 0.$$

Indeed, the existence of such a sequence follows from the fact that the inverse of the mapping $T_0 : U_0/U_0 \cap N \to TU_0$, determined by

$$T_0[x] = Tx,$$

cannot be continuous if TU_0 is not closed. We may clearly assume $\|x_n\| \leq 2$. We distinguish two cases. If $\{x_n\}$ has no convergent subsequence, then it is a closed bounded set and, by hypothesis, $\{Tx_n\}$ is a closed set. By (2.3), $Tx_N = 0$ for some N so that $T_0[x_N] = 0$. Since T_0 is injective, we have $[x_N] = 0$, contradicting (2.3). Thus, $\{x_n\}$ has at least one convergent subsequence, again denoted $\{x_n\}$, $x_n \to x$. By (2.3), $Tx = 0$ so that $[x] = 0$. Thus $[x_n] \to 0$, contradicting (2.3). It follows that TU_0 is closed in Y.

(iii) $\Longrightarrow$ (iv). If TU_0 is closed, then TU - y is a closed and convex subset of Y, and hence weakly closed. In a reflexive space, every weakly closed set has an element of minimal norm and hence (iv) holds. □

There is a useful sufficient condition which ensures that TU is a closed convex subset of Y for U any convex subset of X.

Lemma 2.4. Let X and Y be Banach spaces and let T be a continuous linear mapping of X onto Y. Let U be a convex subset of X such that U + N is closed, where N is the null space of T. Then TU is a closed, convex subset of Y. In particular, if N is finite dimensional and U is a closed flat, then U + N is closed and in this case TU is closed and convex.

Proof: Let U be a convex subset of X and suppose that U + N is closed in X. If J denotes the canonical mapping from X onto X/N then the mapping $T_0 : J(X) \to Y$ defined by

$$T_0 Jx = Tx$$

is linear, continuous and bijective. Now, by the open mapping theorem, the set $J(U+N) = J(U)$ is closed, since the images of U + N and its complement under J are disjoint. Since T_0 is invertible, it follows again from the open mapping theorem that $TU = T_0(J(U))$ is closed in Y. The final statement follows from [2.4, p. 513]. □

There is an analogue result to Theorem 2.2 when the weak-* topology is utilized.

Theorem 2.5. Let X and Y be Banach spaces which are the duals of separable normed linear spaces and let T be a weak-* continuous linear mapping of X onto Y. Let Γ be a weak-* continuous linear mapping of X onto a Banach space Z, let K be a weak-* closed, bounded subset of Z and define $U = \{x \in X : \Gamma x \in K\}$. Then for each $y \in Y$, the minimization problem

$$\alpha = \inf\{\|Tx-y\| : x \in U\}$$

has a solution.

Proof: It clearly suffices to prove the existence of a bounded minimizing sequence. If $\{x_k\}$ is a minimizing sequence, select a bounded set $\{x_k'\} \subset X$ such that $Tx_k' = Tx_k$, $k = 1,2,\ldots$. This is possible since there is a constant $C > 0$ such that

$$\|x\| \leq C\|y\|$$

for every $y \in Y$ and some x satisfying $Tx = y$ [2.8, p. 179]. The remainder proceeds as in the proof of Theorem 2.2. □

We continue by stating a result for unbounded sets $K \subset Z$. We have as an immediate consequence of Lemma 2.4,

<u>Theorem 2.6</u>. Let X and Y be Banach spaces and let T be a continuous linear mapping of X onto Y with finite dimensional null space N. Let $\{Z_\gamma\}$ be a family of Banach spaces and let $\{\Gamma_\gamma\}$ be a family of continuous linear mappings of X into Z_γ. If $\{K_\gamma\}$ is a family of closed convex subsets of Z_γ, and if U is defined by

$$U = \{x \in X : \Gamma_\gamma x \in K_\gamma \text{ for all } \gamma\},$$

then TU is closed in Y if U is contained in an algebraic complement of N.

Proof: By Lemma 2.4 it suffices to show that U + N is closed. This property, however, is a consequence of the fact that the asymptotic cone U_∞ of U satisfies $U_\infty \cap N = \{0\}$; indeed,

$$U_\infty = \bigcap_{\lambda > 0} \lambda(U - u_0)$$

for any $u_0 \in U$ and hence U_∞ is contained in any linear space containing U. The result now follows from a theorem of Dieudonne, applied for the first time in such a context by Atteia [2.2]. □

A final theorem on linear existence follows. Here Ω is a σ-finite measure space.

Theorem 2.7. Let H be a Hilbert space and L a bounded linear operator from H into $L^2(\Omega)$ with closed range. Let N be the null space of L, let $\ell_1,\ldots,\ell_m$ be continuous linear functionals on N, let $F_1,\ldots,F_m$ be $L^1(\Omega)$ functions and $W = \{x \in H : Lx \in L^\infty(\Omega)\}$. For $1 \leq j \leq m$, let L_j be the linear functional on W given by $L_j(x) = \int_\Omega F_j Lx + \ell_j(Px)$ where P is the orthogonal projection of H on N. Let Λ be a compact convex set in $\mathbb{R}^m$ and let

$$U = \{x \in W : \{L_j x\}_1^m \in \Lambda\}.$$

Consider the minimization problem

$$\alpha = \inf\{\|Lx\|_\infty : x \in U\}. \tag{2.4}$$

This minimization problem has a solution.

Proof. Let $H = N \oplus H'$ be the direct sum decomposition of H. L is 1:1 on H' and the range of L on H' coincides with the range of L on H. Further, $x \in W$ if and only if the projection of x onto H' is in W. The range of L is closed in L^2 and hence the range of L on W is closed in L^∞. Suppose $x_k \in U$ and $\|Lx_k\|_\infty \to \alpha$; $x_k = h_k + n_k$ where $n_k \in N$ and $h_k \in H'$. The open mapping theorem implies that $\|h_k\|_H \leq C$ for all k. Let $T : N \to \mathbb{R}^m$ be given by $Tn = (\ell_1(n),\ldots,\ell_m(n))$ and let N_0 be the kernel of T. Then N/N_0 is finite-dimensional and since $\|Tn_k\|$ is bounded there are elements n_k' of N with $n_k - n_k' \in N_0$ and $\|n_k'\|_H \leq C'$ for all k. Let $x_k' = h_k + n_k'$; then $\|x_k'\|_H \leq C + C'$ for all k and $Lx_k' = Lx_k$. Some subsequence of $\{x_k'\}$, again denoted by $\{x_k'\}$, converges weakly to an element x of H. Since x lies in the norm closure of the convex hull of $\{x_k'\}_{k \geq k_0}$ for all k_0, we may assume that x_k' converges in the norm of H to x. Hence, $Lx_k' \to Lx$ in $L^2(\Omega)$. A further subsequence converges a.e. to Lx; hence, $Lx \in L^\infty$ and since $\|Lx_k'\|_\infty \leq \alpha + \epsilon$ for $k \geq k_1(\epsilon)$, we find that $x \in U$ and $\|Lx\|_\infty \leq \alpha$. Thus x is a solution of (2.4). □

2.2 Applications

Example 2.1. Generalized Spline Functions.

Let $X = W^{n,p}(a,b)$, $1 < p \leq \infty$, let $Y = L^p(a,b)$, let $T = D^n + \sum_{j=0}^{n-1} c_j D^j$ where $c_j \in C[a,b]$. Let $\ell_1,\ldots,\ell_m$ be linearly independent elements of X* and define $\Gamma f = \{\ell_j(f)\}_{j=1}^m \in \mathbb{R}^m$. Let K be a convex, compact set in $\mathbb{R}^m$ and set $U = \{f \in X : \Gamma f \in K\}$. For $1 < p < \infty$, it follows directly from Theorem 2.2 that the extremal problem 1.1 has a solution. For $p = \infty$, the result follows from Theorem 2.5; it must of course be observed that the unit ball of $W^{n,\infty}(a,b)$ is weak-* sequentially compact. This latter result is a consequence of the fact that $W_0^{n,\infty}(a,b)$ is the dual of $W_0^{n,1}(a,b)$. Thus, the space $\mathscr{P}_{n-1} \oplus W_0^{n,\infty}(a,b)$, which is equivalent to $W^{n,\infty}(a,b)$, is a dual space of a separable Banach space. Here $\mathscr{P}_{n-1}$ denotes the polynomials of degree n - 1.

Special choices of the functionals $\{\ell_j\}$ such as point evaluations, or, more generally, linear combinations of derivative evaluations through order n - 1, lead to important and interesting problems of approximation theory. The characterization properties of such extremal solutions will be examined in detail in the sequel.

Example 2.2. Bang-Bang Solutions for Elliptic Problems

Let Ω be a bounded, open subset of $\mathbb{R}^m$, let L be a continuous, bijective mapping of $W^{2k}(\Omega) \cap W_0^k(\Omega)$ onto $L^2(\Omega)$. Let X be the vector subspace of $W^{2k}(\Omega) \cap W_0^k(\Omega)$, consisting of all f with $Lf \in L^\infty(\Omega)$, and set $Y = L^\infty(\Omega)$ with T the restriction of L to X. Let $\ell_1,\ldots,\ell_n$ be n linearly independent linear functionals on X defined by

$$\ell_j(f) = \int_\Omega g_j Lf$$

for $g_1,\dots,g_n$ in $L^1(\Omega)$. Then an application of Corollary 1.3 or Theorem 2.7 shows that the minimization problem (1.1) has a solution, where U is defined by,

$$U = \{f \in X : \{\ell_j f\}_{j=1}^n \in \Lambda\},$$

where Λ is a compact convex subset of $\mathbb{R}^n$. In §4, we shall investigate the bang-bang properties of these solutions.

Example 2.3. The Intersection of Half-Space Constraints

Let X and Y be real Banach spaces and let T be a continuous linear mapping of X onto Y. Let $\{L_\gamma\}_{\gamma \in \Gamma}$ be a family of real continuous linear functionals on X and let $r = (r_\gamma)_{\gamma \in \Gamma}$ be prescribed. Define

$$U = \{x \in X : L_\gamma x \geq r_\gamma, \ \gamma \in \Gamma\}.$$

Suppose $\{x \in X : L_\gamma x \geq 0 \text{ for each } \gamma\} \cap N = \{0\}$. Then, if U is non-empty, TU is closed and convex in Y as is seen by a direct application of the proof of Theorem 2.6. If Y is reflexive, then by Theorem 2.3 the minimization problem (1.1) has a solution.

Example 2.4. Interpolation on the Integers

Let $X = W^{n,p}(\mathbb{R})$, $Y = L^p(\mathbb{R})$ and $T = D^n$. Let F and G be two fixed functions in X with $F \leq G$. Then, if $U = \{f \in X : F(i) \leq f(i) \leq G(i) \text{ for all integers } i\}$ it follows that U is weakly closed if $1 < p < \infty$ and weak-* closed if $p = \infty$. Then, Theorem 2.3 for $1 < p < \infty$ and Theorem 2.5 for $p = \infty$ imply that (1.1) has a solution.

Remarks. Lemma 2.4 and the implication (2.1i) ⟹ (2.1ii) were proved in [1.5]. Example 2.3 is proved also in [1.5]. The implication (2.1ii) ⟹ (2.1iii) is proved in Goldberg [2.5, p. 110]. A host of papers dealing with the minimization problem (2.1) has appeared in the literature. No doubt influenced by the early fundamental paper of Golomb and Weinberger [2.6], the French school led the way in the investigation of the abstract formulation of the problem, with the goal of application to generalized spline functions. The papers of Anselone and Laurent [2.1], Atteia [2.2] and Aubin [2.3] are representative. A more complete bibliography, at least in relation to spline function applications, may be found in Jerome and Varga [2.7]. However, we do emphasize that our present applications go beyond this, including multivariate problems, for example (cf. example 2.2).

REFERENCES

2.1 P. M. Anselone and P. J. Laurent, "A general method for the construction of interpolating or smoothing splines," Numer. Math., 12 (1968), 66-82.

2.2 M. Atteia, "Fonctions splines defines sur un ensemble convexe," Ibid., 12 (1968), 192-210.

2.3 J. P. Aubin, "Interpolation et approximation optimales et 'Spline functions'," J. Math. Anal. Appl., 24 (1968), 1-24.

2.4 N. Dunford and J. T. Schwartz, Linear Operators, Part I, Wiley Interscience, 1957.

2.5 S. Goldberg, Unbounded Linear Operators, McGraw-Hill, New York, 1966.

2.6 M. Golomb and H. F. Weinberger, "Optimal approximation and error bounds," in, On Numerical Approximation, (R. E. Langer, ed.), Univ. of Wisconsin Press, Madison, Wis., 1959, pp. 117-190.

2.7 J. Jerome and R. S. Varga, "Generalizations of spline functions and applications to nonlinear boundary value and eigenvalue problems," in, Theory and Applications of Spline Functions, (T. N. E. Greville, editor), Academic Press, New York, 1969, pp. 103-155.

2.8 A. E. Taylor, Introduction to Functional Analysis, Wiley, New York, 1958.

PART II: CHARACTERIZATION THEOREMS

§3. Nonlinear Operators in L^p, $1 < p \leq \infty$.

3.1 General Theorems

In this section, we shall use the Frechet derivative to characterize solutions of the problem (1.1)

Theorem 3.1. Let X be a Banach space, T a continuous linear mapping of X into $L^p(a,b)$, $1 < p < \infty$, let $\ell_1,\ldots,\ell_m$ be linearly independent elements of X^*, let K be a closed set in $\mathbb{R}^m$ and set

$$\Gamma(x) = (\ell_1(x),\ldots,\ell_m(x)) \in \mathbb{R}^m.$$

Define

$$U = \{x \in X : \Gamma x \in K\},\ U_0 = \{x \in X : \Gamma x = 0\}$$

and suppose TU_0 is closed in $L^p(a,b)$. Suppose x_0 is a solution of the minimization problem

$$\alpha = \inf\{\|Tx\| : x \in U\}.$$

Then there is a function $g \in L^q$, where $1/p + 1/q = 1$, such that

(3.1)

$$\text{(i)}\quad \int_a^b gTx = 0 \text{ for all } x \in U_0$$

$$\text{(ii)}\quad gTx_0 \geq 0 \text{ a.e. on } (a,b)$$

$$\text{(iii)}\quad |g| = \alpha^{(1-p)}|Tx_0|^{p-1} \text{ a.e.}$$

$$\text{(iv)}\quad \alpha = \int_a^b gTx_0$$

We shall not prove this theorem since it is an immediate consequence of Theorem 3.2.

Theorem 3.2. Let T be a continuous Frechet differentiable mapping from a Banach space X into $L^p(\Omega)$ where Ω is a measurable subset of Euclidean space, $1 < p < \infty$. Let Γ be a continuous linear operator from X into a Banach space Z; let K be a closed convex subset of Z and let

$$U = \{x \in X : \Gamma x \in K\},$$

$$U_0 = \{x \in X : \Gamma x = 0\}.$$

Suppose x_0 is a solution of the minimization problem

$$\alpha = \inf\{\|Tx\| : x \in U\}.$$

Let L be the Frechet derivative of T at x_0. Then, provided LU_0 is closed, there is a function $g \in L^q$, where $1/p + 1/q = 1$, such that

$$\text{(3.2)} \quad \begin{aligned} &\text{(i)} && \int_\Omega gLx = 0 \text{ for all } x \in U_0 \\ &\text{(ii)} && gTx_0 \geq 0 \text{ a.e. on } \Omega \\ &\text{(iii)} && |g| = \alpha^{(1-p)}|Tx_0|^{p-1} \\ &\text{(iv)} && \alpha = \int_\Omega gTx_0 \end{aligned}$$

Proof. We know that $\alpha = \|Tx_0\| \leq \|T(x_0+\epsilon v)\|$ for all $v \in U_0$. Suppose there is a $v \in U_0$ for which

$$\|Tx_0+Lv\| \leq \alpha - \delta$$

for some $\delta > 0$. Then

$$\alpha \le \|T(x_0+\epsilon v)\| = \|Tx_0+\epsilon Lv\| + o(\epsilon)$$

$$\le (1-\epsilon)\|Tx_0\| + \epsilon\|Tx_0+Lv\| + o(\epsilon)$$

$$\le \alpha - \epsilon\delta + o(\epsilon) < \alpha$$

when ϵ is sufficiently small. Thus

$$\alpha = \inf\{\|Tx_0+Lv\| : v \in U_0\}$$

This **infimum** is the norm of the coset $Tx_0 + LU_0$ in the quotient space L^p/LU_0 and this in turn is just the norm of Tx_0 as a linear functional on $S = \{g \in L^q : \int_\Omega gLv = 0 \text{ for all } v \in U_0\}$ [2.8, p. 227, prob. 5, 6]. Thus,

$$\alpha = \sup\{|\int_\Omega (Tx_0)g| : \|g\|_q \le 1,\ g \in S\}$$

and this supremum is actually a maximum since Tx_0 is a weakly continuous functional on the weakly compact unit ball in S. The remainder of the conclusions follow from equality in Hölder's inequality [3.9, p. 34]. □

Remark. If L is a continuous linear mapping of X into $L^p(a,b)$ with closed range and satisfies property (D) with respect to strong convergence then LU_0 is closed by Theorem 2.3. In particular, this holds if the null space of L is finite dimensional.

Theorem 3.3. Suppose T is a continuous Frechet differentiable operator mapping a Banach space X into $L^\infty(\Omega,\mu)$ where Ω is a σ-finite measure space. Let $\ell_1,\dots,\ell_m$ be linearly independent elements of X^*, let K be a compact convex subset of $\mathbb{R}^m$ and define $\Gamma x = (\ell_1(x),\dots,\ell_m(x))$ and

$$U = \{x \in X : \Gamma x \in K\},$$

$$U_0 = \{x \in X : \Gamma x = 0\}.$$

Suppose x_0 is a solution of the variational problem

$$\alpha = \inf\{\|Tu\|_{L^\infty} : u \in U\}$$

Let L be the Frechet derivative of T at x_0. If LU_0 is weak-* closed in $L^\infty(\Omega,\mu)$ and if the null space of the adjoint L^a is finite dimensional then there is an $h \in L^1(\Omega,\mu)$ with

(i) $\int_\Omega |h| d\mu = 1$

(ii) $0 = \int_\Omega hLvd\mu$ for all $v \in U_0$

(3.3) (iii) $hTx_0 \geq 0$ a.e. μ

(iv) $|Tx_0| = \alpha$ a.e. μ where $h \neq 0$.

(v) $\alpha = \int_\Omega hTx_0 d\mu$

Proof. As in Theorem 3.2 we find that

$$\alpha = \inf\{\|Tx_0+Lv\| : v \in U_0\}$$

and hence α is the coset norm of $Tx_0 + LU_0$ in L^∞/LU_0. Because LU_0 is weak-* closed, the quotient space L^∞/LU_0 is the dual space of $S = \{f \in L^1(\Omega,\mu) : 0 = \int_\Omega fLv$ for all $v \in U_0\}$. Hence, α is the norm of Tx_0 as a linear functional on S. The assumption that the null space of L^a is finite-dimensional implies that S is finite-dimensional. Hence, there is an $h \in S$ with $\|h\|_1 = 1$ and $\alpha = \int_\Omega hTx_0$. The latter yields conclusions (iii) and (iv) since we have equality in

Hölder's inequality. □

Remarks. The assertion that L^∞/LU_0 is the dual of S in the proof of Theorem 3.3 requires the following argument. Indeed, what is true [2.8, p. 188] is that L^∞/S^0 is the dual of S, where

$$S^0 = \{x^* \in L^\infty(\Omega,\mu) : x^*(x) = 0 \text{ for all } x \in S\}.$$

The result follows if it can be shown that $S^0 = LU_0$. Now, clearly, $S^0 = ({}^0LU_0)^0$ where, for a subspace $M \subset L^\infty(\Omega,\mu)$,

$${}^0M = \{x \in L^1(\Omega,\mu) : x'(x) = 0 \text{ for all } x' \in M\}.$$

In general, $({}^0M)^0 \neq M$ for a closed subspace of $L^\infty(\Omega,\mu)$; equality holds if and only if M is weak-* closed [2.8, p. 232]. The hypothesis that LU_0 is weak-* closed then guarantees the result.

Corollary 3.4. Suppose the unit ball of X is weakly sequentially compact (or weak-* sequentially compact) and that the range of L has finite codimension in L^∞ and L sends weakly convergent (weak-* convergent) sequences in X to weak-* convergent sequence in L^∞. Then LU_0 is weak-* closed and the null space of L^a is finite dimensional and hence the conclusions of Theorem 3.3 hold.

Proof. Since the range of L has finite codimension in L^∞ it is closed and the dimension of the null space of L^a is just the codimension of L(X) in L^∞. To prove LU_0 is weak-* closed it suffices to prove that its unit ball is weak-* sequentially closed [2.4, p. 426]. Suppose then that $x_n \in U_0$ and Lx_n converges weak-* to u, $\|Lx_n\|_\infty \leq 1$. Note that X/U_0 is finite dimensional and hence so is $L(X)/LU_0$. Hence, LU_0 is closed in L(X) and therefore is closed in L^∞ and is a Banach space. By the open mapping theorem we may assume that

$\|x_n\| \leq C$ for some C and all n. Thus some subsequence $\{x_{n_j}\}$ converges weakly (converges weak-$*$) to an element $x \in U_0$; further, Lx_{n_j} converges weak-$*$ to Lx. But Lx_{n_j} also converges weak-$*$ to u so $u = Lx$ and the corollary is proved. $\square$

3.2 Applications

In this section, we analyze in detail several of the examples of §1.

<u>Theorem 3.5</u>. Let ψ be a C^1 function on $[0,1] \times \mathbb{R}^m$, let $1 < p \leq \infty$ and let $g \in L^p(0,1)$. Let U be a flat in $W^{n,p}(0,1)$ of codimension n, with parallel subspace U_0, defined by n linear functionals continuous on $C^{n-1}[0,1]$ and linearly independent over $\mathscr{P}_{n-1}$. Then the boundary value problem

$$\text{(3.4)} \quad \begin{aligned} &\text{(i)} \quad y^{(n)}(t) + \psi(t,y(t),y'(t),\ldots,y^{(n-1)}(t)) = Ty(t) = g(t) \\ &\text{(ii)} \quad y \in U \end{aligned}$$

has a solution $y \in W^{n,p}(0,1)$ provided the minimization problem

$$\text{(3.5)} \qquad \alpha = \inf\{\|Tf-g\| : f \in U\}$$

has a solution y_0 and provided the linearized problem

$$\text{(3.6)} \quad \begin{aligned} &\text{(i)} \quad y^{(n)}(t) + \sum_{j=1}^{n} \psi_j(t,y_0(t),\ldots,y_0^{(n-1)}(t))y^{(j-1)}(t) \\ &\qquad = L_{y_0}y(t) = h(t) \\ &\text{(ii)} \quad y \in U_0 \end{aligned}$$

has a solution for each $h \in L^p(0,1)$. In particular, (3.5) has a solution if ψ satisfies the growth condition

$$(3.7) \qquad |\psi(t,x)| \leq M + \delta \max_{1 \leq i \leq n} |x_i|, \text{ for all } t \in [0,1]$$

$$\text{and } x = (x_1,\ldots,x_m) \in \mathbb{R}^m$$

where δ is a nonnegative constant satisfying $\delta\lambda < 1$. Here
$\lambda = \sup_{\substack{0 \leq t \leq 1 \\ 0 \leq j \leq n-1}} (\int_0^1 |K^{(j)}(t,\xi)|^q d\xi)^{1/q}$ and $K(\cdot,\cdot)$ is the kernel of
the inverse of $D^n : U_0/U_0 \cap \mathscr{P}_{n-1} \to L^p(0,1)$. Here, differentiation is taken with respect to the first argument of K.

Proof. We view T as an operator from $W^{n,p}(0,1)$ into $L^p(0,1)$. Suppose (3.7) holds. Theorem 1.1 is applicable for $1 < p < \infty$ and Theorem 1.2 for $p = \infty$. Note that, for $1 < p < \infty$, $W^{n,p}(0,1)$ and $L^p(0,1)$ are reflexive, and, for $p = \infty$, $W^{n,\infty}(0,1)$ and $L^\infty(0,1)$ may be viewed as the duals of $W^{n,1}(0,1)$ and $L^1(0,1)$. U is clearly weakly closed if $1 < p < \infty$ and weak-$*$ closed if $p = \infty$ since both weak and weak-$*$ convergence in U imply uniform convergence of derivatives through order $n - 1$. This also implies the weak and weak-$*$ continuity of T. Thus, to conclude the existence of a solution of (3.5), we verify that minimizing sequences are bounded. Thus, if

$$(3.8) \qquad \|y^{(n)}+\psi(\cdot,y,\ldots,y^{n-1})\| \leq C, \; y \in U,$$

as is the case if y belongs to a minimizing sequence, then by (3.7),

$$(3.9) \qquad \|y^{(n)}\| \leq C + M + \delta \max_{1 \leq i \leq n} \|y^{(i-1)}\|$$

where norms are taken in $L^p(0,1)$. One sees immediately, using the representation,

$$y(t) = P_{n-1}(t) + \int_0^1 K(t,\xi)y^{(n)}(\xi)d\xi,$$

and the properties $y - P_{n-1} \in U_0$, and the finite dimensionality of $\mathscr{P}_{n-1}$, that there is a constant $0 < C_1$ such that

$$(3.10)\quad \|y^{(i-1)}\| \leq C_1 + \max_{0 \leq t \leq 1} \left(\int_0^1 |K^{(i-1)}(t,\xi)|^q d\xi\right)^{1/q} \|y^{(n)}\|.$$

(3.9) and (3.10), together with the hypothesis on δ, imply that $\|y^{(n)}\| \leq C_2$, independent of y satisfying (3.8). It follows that (3.5) has a solution if the growth condition (3.7) holds for ψ.

Now suppose that (3.5) has a solution y_0 and that the linearized equation (3.6) has a solution for each $h \in L^p(0,1)$. Then the Frechet derivative of T at y_0 is the operator L defined by the left hand side of (3.6i). For $1 < p < \infty$, the hypotheses of Theorem 3.2 are satisfied and for $p = \infty$ the hypotheses of Theorem 3.3 are satisfied. In both cases, L maps U_0 onto $L^p(a,b)$ and hence the function g in (3.2i) and the function h in (3.3ii) are identically 0. Thus $\alpha = 0$ by (3.2iv) and (3.3v). $\square$

<u>Theorem 3.6</u>. Consider the nonlinear Dirichlet problem defined by (1.9) where Ω, T, L and ψ satisfy the hypotheses stated in Example 1.4. Then (1.9) has a solution, provided the linearized problem

$$(3.11)\quad \begin{aligned} &(i)\quad Ly + \sum_{|\gamma| < \nu} \psi_\gamma(\cdot, D^\beta y_0) D^\gamma y = h \\ &(ii)\quad y \in W^{2k,2}(\Omega) \cap W_0^{k,2}(\Omega) \end{aligned}$$

has a solution for each $h \in L^2(\Omega)$. Here β ranges over all indices of length less than ν in (3.11i).

Proof. It was demonstrated in Example 1.4 that the corresponding minimization problem

$$\alpha = \inf\{\|Ty-g\| : y \in U\}$$

has a solution. It remains to show $\alpha = 0$. The Frechet derivative of T is simply the operator defining the left hand side of (3.11i) and the hypothesis that the linearized problem is solvable for each $h \in L^2(\Omega)$ implies that $g \equiv 0$ in (3.2i). Thus $\alpha = 0$ from (3.2iv). □

We shall close section 3.2 by giving the explicit solution of the minimization problem posed in Example 1.6. We recall that $X = W^{1,\infty}(0,1)$, $y = L^{\infty}(0,1)$ and $Tf(t) = f'(t)\exp(f(t))$, $0 \leq t \leq 1$. It was shown in Example 1.6 that a solution f_0 exists satisfying

$$\|Tf_0\| = \alpha = \inf\{\|Tf\| : f \in W^{1,\infty}(0,1),\ f(0) = 0,\ f(1) = 1\}.$$

Now the Frechet derivative of T at f_0 is

$$Lg = e^{f_0}g' + f_0'e^{f_0}g = A'g + Ag' = (Ag)'$$

where $A = e^{f_0}$. According to Theorem 3.3, there is an $h \in L^1(0,1)$, $\|h\| = 1$, with

$$\int_0^1 h(t)Lg(t)\,dt = 0 \tag{3.12}$$

for all $g \in W^{1,\infty}(0,1)$ satisfying $g(0) = g(1) = 0$; also, $|Tf_0| = \alpha$ when $h \neq 0$ together with $(Tf_0)h \geq 0$. It follows immediately from (3.12) and elementary distribution theory [3.2] that h is constant and hence $Tf_0 \equiv \alpha$ or $Tf_0 \equiv -\alpha$ on (0,1). Thus, integrating and employing the initial conditions, we obtain

$$f_0(t) = \ln[(e-1)t+1], \ 0 \le t \le 1,$$

so that $\alpha = e - 1$ and $Tf_0 \equiv \alpha$.

3.3 The L^p-Splines, $1 < p < \infty$.

In this section we consider in detail the solutions described in Example 2.1 for $1 < p < \infty$. We make, as special choice of the linear functionals, the extended-Hermite-Birkhoff functionals consisting of linear combinations of derivatives. A much more general result follows from Theorem 3.1, however. Consider, then, a mesh

$$a = x_0 < x_1 < \dots < x_m = b$$

of $[a,b]$ and, associated with each of the points x_i, consider the continuous linear functionals L_{ij} on $W^{n,p}(a,b)$ defined by

$$(3.13) \quad L_{ij}f = \sum_{\nu=0}^{n-1} a_{ij}^{(\nu)} f^{(\nu)}(x_i), \ j = 0,\dots,k_i-1, \ i = 0,\dots,m,$$

for prescribed real numbers $a_{ij}^{(\nu)}$ such that, for each i, the k_i n-tuples $(a_{ij}^{(0)},\dots,a_{ij}^{(n-1)})$ are linearly independent; here $1 \le k_i \le n$ for $i = 0,\dots,m$. Let L be given as in Example 2.1. Let I_{ij}, $j = 0,\dots,k_i-1$, $i = 0,\dots,m$ be prescribed real closed intervals. Consider, for $1 < p < \infty$, the minimization problem

$$(3.14) \quad \|Ls\| = \alpha = \inf\{\|Lf\| : L_{ij}f \in I_{ij}, \ j = 0, \ k_i-1, \ i = 0,\dots,m\}.$$

As remarked, solutions exist for (3.14) and Ls is unique by the strict convexity of $L^p(a,b)$, $1 < p < \infty$. We wish to characterize these solutions. We require some preliminary notation.

Now for each fixed $i = 0,\dots,m$ let A_i be the $k_i \times n$ matrix

$$A_i = (a_{ij}^{(\nu)})$$

where j denotes row and ν column indices. Let $\hat{A}_i$ be any nonsingular $n \times n$ augmentation of A_i. Let $H_i = (h_{ij}^{(\nu)})$ be the inverse of the transpose of $\hat{A}_i$. If operators σ_ν are defined on suitably smooth functions by

$$(3.15) \qquad \sigma_\nu f = \sum_{j=0}^{n-\nu-1} (-1)^{j+1}[(c_{j+\nu+1})|Lf|^{p-1} \text{ signum } Lf]^{(j)}$$

for $\nu = 0,\ldots,n-1$ and if operators R_{ij} are defined by

$$(3.16) \qquad R_{ij} = \sum_{\nu=0}^{n-1} h_{ij}^{(\nu)}\sigma_\nu, \; j = 0,\ldots,n-1, \; i = 0,\ldots,m$$

then [cf., 3.6, Lemma 3.1], if the notation $[\cdot]_i$ is defined by

$$[\varphi]_i = \varphi(x_i+) - \varphi(x_i-), \text{ for } 0 < i < m,$$

and

$$[\varphi]_0 = \varphi(x_0+), \; [\varphi]_m = -\varphi(x_m-),$$

we have, for $i = 0,1,\ldots,m$,

$$(3.17) \qquad \sum_{i=0}^{m}\sum_{j=0}^{n-1} D^j g(x_i)[\sigma_j f]_i = \sum_{i=0}^{m}\sum_{j=0}^{n-1} L_{ij} g[R_{ij} f]_i$$
$$= \int_a^b \{[|Lf|^{p-1}\text{signum } Lf]Lg\}$$
$$- \int_a^b \{L^*[|Lf|^{p-1}\text{signum } Lf]g\}$$

for all f, g for which (3.17) is meaningful. Notice that $\hat{A}_i$ induces operators L_{ij} for $k_i \le j \le n-1$.

Theorem 3.7. Suppose $c_j \in C^n[a,b]$ and $1 < p < \infty$. Then s is a solution of the minimization problem (3.14) if and only if $s \in W^{n,p}(a,b)$ and

$$\text{(i)}\quad L^*|Ls|^{p-1}\text{signum } Ls = 0 \text{ on } (x_i,x_{i+1}),\ i = 0,\dots,m-1,$$

$$\text{(ii)}\quad L_{ij}s \in I_{ij},\ j = 0,\dots,k_i - 1,\ i = 0,\dots,m$$

$$(3.18)\quad \text{(iii)}\quad [R_{ij}s]_i = 0 \text{ if } j = k_i,\dots,n-1,\ i = 0,\dots,m$$

$$\text{(iv)}\quad [R_{ij}s]_i \leq 0 \text{ if } L_{ij}s > \underline{r}_{ij},\ j = 0,\dots,k_i - 1,\ i = 0,\dots,m$$

$$\text{(v)}\quad [R_{ij}s]_i \geq 0 \text{ if } L_{ij}s < \overline{r}_{ij},$$

where $I_{ij} = [\underline{r}_{ij},\overline{r}_{ij}]$.

Proof. $\theta(f) = \|Lf\|$ is a convex functional on the convex set U defined by (3.18ii). It is known [3.8, Theorem 2.1] that Ls is the unique solution of (3.14) if and only if

$$(3.19)\qquad \int_a^b \{|Ls|^{p-1} \text{ signum } Ls\}Le\,dt \geq 0$$

for all e in the support cone $\{\lambda(f-s) : \lambda > 0,\ f \in U\}$. Now, if (3.18) holds, then the integration by parts formula (3.17) implies that (3.19) holds and hence that s solves (3.14). Conversely, the direct implications utilize the theory of distributions [3.2] as well as (3.17). The details are exactly as in the proof of the special case p = 2 carried out in [3.6, Theorem 7.2]. □

Corollary 3.8. Let $L = D^n$ and suppose that the functionals L_{ij} consist of consecutive derivative evaluations of the form

$$L_{ij}f = f^{(j)}(x_i)$$

where j is some integer satisfying $0 \leq j \leq k_i - 1$ and $0 \leq i \leq m$. Then s is the solution of (3.14) if and only if (3.18ii) holds and

(i) $D^n[|D^n s|^{p-1} \text{ signum } D^n s](x) = 0,\ x \neq x_i,$

(ii) $|D^n s|^{p-1}$ signum $D^n s$ is in the class C^{n-k_i-1} in a neighborhood of x_i, $i = 0,\ldots,m,$

(3.20) (iii) $(-1)^{n-j-1}[D^{n-j}|D^n s|^{p-1} \text{ signum } D^n s]_i \geq 0$ if $s^{(j-1)}(x_i) < \overline{r}_{i,j-1},\ j = 0,\ldots,k_i - 1,\ i = 0,\ldots,m,$

(iv) $(-1)^{n-j-1}[D^{n-j}|D^n s|^{p-1} \text{ signum } D^n s]_i \leq 0$ if $s^{(j-1)}(x_i) > \underline{r}_{i,j-1},\ j = 0,\ldots,k_i - 1,\ i = 0,\ldots,m.$

In particular, if all of the constraint intervals reduce to points then s is characterized by (3.20i, ii).

<u>Remarks</u>. Theorem 3.1 is a well-known result in duality theory (cf. Holmes [3.3]). The generalization contained in Theorem 3.2 as well as Theorem 3.3 are probably new in their specific formulations, although the underlying ideas are totally familiar. The result of Theorem 3.5 is a generalization of a result of Fisher [1.4], where initial value problems were considered. In this case, of course, the linearized problem always possesses a solution. The result of Theorem 3.6 may be paraphrased to assert that, if the linearized Dirichlet problem is solvable, then so is the corresponding nonlinear problem.

The delineation of the structure of the L^p-splines in section 3.3 was first carried out by Golomb [3.1] for $L = D^n$ and for equality constraints and achieved for general L and inequality constraints by Jerome [3.4] by variational methods (cf. also Mangasarian and

Schumaker [3.7]). [3.4] includes characterizations for nonlinear operators as well. The existence of a solution of (3.14) for point evaluations by direct variational methods was first demonstrated by Jerome and Schumaker [3.5]. The case of Theorem 3.7 for p = 2 was proved by Jerome and Schumaker [3.6, Theorem 7.2]. The proof of the result for $p \neq 2$ simply substitutes $|Ls|^{p-1}$ signum Ls for Ls.

REFERENCES

3.1 M. Golomb, "$H^{m,p}$ extensions by $H^{m,p}$-splines," J. Approximation Theory, 5 (1972), 238-275.

3.2 I. Halperin, Theory of Distributions, University of Toronto Press, Toronto, 1952.

3.3 R. Holmes, "A Course on Optimization and Best Approximation," Springer-Verlag Lecture Series in Mathematics, New York, 1972.

3.4 J. W. Jerome, "Linearization in certain nonconvex minimization problems and generalized spline projections," in, Spline Functions and Approximation Theory, Birkhäuser-Verlag, Basel, 1973.

3.5 J. W. Jerome and L. L. Schumaker, "Characterizations of functions with higher order derivatives in L_p," Trans. Amer. Math. Soc., 143 (1969), 363-371.

3.6 ________, "On L-g splines," J. Approximation Theory, 2 (1969), 29-49.

3.7 O. L. Mangasarian and L. L. Schumaker, "Splines via optimal control," in, Approximations With Special Emphasis on Spline Functions, Academic Press, New York, 1969, pp. 119-156.

3.8 B. N. Pshenichnyi, Necessary Conditions for an Extremum, Marcel Dekker, Inc., New York, 1971.

3.9 K. Yosida, Functional Analysis, Academic Press, New York, 1965.

§4. L^∞ Minimization Problems for Elliptic Operators

In this section, we examine in detail the bang-bang properties of the solutions of the problem considered in Example 2.2.

Let Ω be a bounded domain in R^n $(n \geq 2)$ and let L be a linear partial differential operator on Ω of order 2m. Our basic assumption in this section is that the generalized Dirichlet problem has a unique smooth solution; that is, for each $f \in L^2(\Omega)$, there is a unique $u \in W^{2m}(\Omega) \cap W_0^m(\Omega)$ with $Lu = f$, in the sense of distributions. This will be true, for example, if the boundary of Ω is of class C^{2m} and if L satisfies conditions A_1, A_2, and B^m of [4.2, p. 33, 51] for the operator $L + \lambda I$ when λ is sufficiently large.

Let $F_1, \ldots, F_N$ be real integrable functions on Ω which are continuous on Ω except at possibly a finite number of points and which are linearly independent over any set of positive measure in Ω; that is, if $\Sigma_{1 \leq i \leq N}\, c_i F_i = 0$ on a set of positive measure, then $c_1 = \ldots = c_N = 0$. This condition will be satisfied, for example, if $F_1, \ldots, F_N$ are real analytic on Ω except at a finite number of points. Let $\underline{r} \in R^N$ and let

$U = U(\underline{r}) = \{u \in W^{2m}(\Omega) \cap W_0^m(\Omega) : \int_\Omega LuF_i = r_i,\ i = 1, \ldots, N$ and $Lu \in L^\infty(\Omega)\}$.

Theorem 4.1. Let $\psi \in L^\infty(\Omega)$. The minimization problem

$$\alpha = \inf\{\|Lu-\psi\|_{L^\infty} : u \in U\} \tag{4.1}$$

has a unique solution $\varphi \in U$; the difference $L\varphi - \psi$ has constant modulus on Ω. Furthermore, there are disjoint open sets $\Omega+$ and $\Omega-$, whose dense union omits only a set of measure zero in Ω, such that $L\varphi - \psi = \alpha$ on $\Omega+$ and $L\varphi - \psi = -\alpha$ on $\Omega-$.

Proof: Solutions of (4.1) were shown to exist in Example 2.2. Thus, let φ be any solution of (4.1); suppose that $|L\varphi-\psi| \leq \alpha - \delta$, $\delta > 0$, on a set B of positive measure in Ω. Let $A = \Omega - B$ and by taking a subset of B if necessary, assume that A also has positive measure. By assumption, there is a $g \in W^{2m}(\Omega) \cap W_0^m(\Omega)$ with

$$Lg = \begin{cases} L\varphi & \text{on } A \\ \psi & \text{on } B. \end{cases}$$

We claim that there is an $h \in W^{2m}(\Omega) \cap W_0^m(\Omega)$ with $Lh = \psi$ on A, $Lh \in L^\infty(\Omega)$, and $\int_\Omega (Lh)F_i = \int_\Omega (Lg)F_i$ for $i = 1,\ldots,N$. To see this, let C consist of all functions $h \in W^{2m}(\Omega) \cap W_0^m(\Omega)$ with $Lh = \psi$ on A and $Lh \in L^\infty(\Omega)$. C is convex; let Λ be defined by

$$\Lambda(u) = \{\int_\Omega (Lu)F_i\}_{i=1}^N, \quad u \in W^{2m}(\Omega) \cap W_0^m(\Omega)$$

If the claim were false, the vector $\Lambda(g)$ could be separated by a hyperplane from $\Lambda(C)$. Thus there would be scalars $c_1,\ldots,c_N$, not all zero, with

$$\sum_i^N c_i \int_\Omega (Lh)F_i \leq \sum_1^N c_i \int_\Omega (Lg)F_i, \text{ for all } h \in C.$$

Hence, if we let $F = \Sigma_{1 \leq i \leq N}\, c_iF_i$, then

$$\int_A \psi F + \int_B (Lh)F \leq \int_\Omega (Lg)F.$$

Lh can be any L^∞ function on B; this clearly implies that $F = 0$ on B. Since $F_1,\ldots,F_N$ are linearly independent over B, we have $c_1 = \ldots = c_N = 0$. Thus such an h exists.

Let $G = \varphi - \varepsilon(g-h)$. Then $G \in U$ and

$$LG - \psi = \begin{cases} (1-\varepsilon)(L\varphi-\psi) & \text{on } A \\ L\varphi - \psi + \varepsilon(Lh-\psi) & \text{on } B \end{cases}$$

and so $|LG-\psi| \leq \alpha - \delta'$, $\delta' > 0$, on all of Ω when ε is sufficiently small, contradicting the definition of α. Hence, if φ is a solution of (1.1), then $|L\varphi-\psi| = \alpha$ a.e. on Ω.

Suppose φ_1 is another solution of (1.1), $\varphi_1 \in W^{2m}(\Omega) \cap W_0^m(\Omega)$. Then $(1/2)(\varphi+\varphi_1)$ is also a solution so that $|L(\varphi+\varphi_1)-2\psi| = 2\alpha$ a.e. on Ω. Hence, $L\varphi - \psi = L\varphi_1 - \psi$ a.e. on Ω which implies $\varphi = \varphi_1$ by the uniqueness of the solution to the generalized Dirichlet problem.

Finally, let φ be the solution of (1.1); let $P = \{x \in \Omega : L\varphi - \psi = \alpha$ and $N = \{x \in \Omega : L\varphi - \psi = -\alpha\}$. Suppose that the closure of P meets the closure of N in a set S of positive measure. Then we shall show that there is a solution G of (1.1) with $\|LG-\psi\| < \alpha$, a contradiction. Let $C = \{v \in W^{2m}(\Omega) \cap W_0^m(\Omega) : Lv \geq \psi$ on N and $Lv \leq \psi$ on P$\}$. We claim there is a $v \in C$ with $\int LvF_i = \int L\varphi F_i$ for $i = 1,\ldots,N$. If not, since C is convex there are scalars $c_1,\ldots,c_N$ not all zero with

$$\int_\Omega Lv(\Sigma\, c_iF_i) \leq \int_\Omega L\varphi(\Sigma\, c_iF_i),\quad v \in C.$$

Hence, $F = \Sigma\, c_iF_i$ is negative on N and positive on P and hence by continuity vanishes on S except for a finite point set. Since S has positive measure, $c_1 = \ldots = c_N = 0$. Hence, such a v exists. Let $G = (1-\varepsilon)\varphi + \varepsilon v$. Then $G \in U$ and

$$LG - \psi = \begin{cases} (1-\varepsilon)\alpha + \varepsilon(Lv-\psi) & \text{on } P \\ -(1-\varepsilon)\alpha + \varepsilon(Lv-\psi) & \text{on } N. \end{cases}$$

Thus $|LG-\psi| \leq \alpha - \delta'$ a.e. on Ω, in contradiction to the choice of φ.

Hence, if P and N both have positive measure, then the closures of P and N have only a set of measure zero in common. Let $\Omega_1 = (\Omega-\overline{P}) \cup (\Omega-\overline{N})$; then Ω_1 is open, $\Omega - \Omega_1$ has measure zero, and $L\varphi - \psi = \alpha$ on the open set $\Omega - \overline{N}$, $L\varphi - \psi = -\alpha$ on the open set $\Omega - \overline{P}$. $\square$

Corollary 4.2. Let $L = \Sigma_{|\alpha|,|\beta| \leq m} (-1)^{|\alpha|} D^{\alpha}(c_{\alpha\beta}D^{\beta})$ be strongly elliptic in Ω; let φ be the solution of (4.1). Then there is a dense open set Ω_1 of Ω where $\Omega \backslash \Omega_1$ has measure zero such that

(i) if $c_{\alpha\beta} \in C^{k+|\alpha|}(\Omega) \cap C^m(\Omega)$ and $\psi \in C^k(\Omega)$, then $\varphi \in C^h(\Omega_1)$ where $h = 2m + k - [n/2] - 1$,

(ii) if $c_{\alpha\beta} \in C^{\infty}$ for all α, β and $\psi \in C^{\infty}$, then $\varphi \in C^{\infty}(\Omega_1)$,

(iii) if $c_{\alpha\beta}$ is real analytic on Ω for all α, β and if ψ is real analytic on Ω, then φ is real analytic on Ω_1.

Proof. The set Ω_1 is described above. Assertion (ii) follows from (i); (i) and (iii) may be found in [4.2], pages 56 and 205, respectively. □

Remark (1). Conclusion (iii) actually holds under the weaker assumption on L of ellipticity.

Remark (2). Let L be a bijective continuous mapping of $W^{2m}(\Omega) \cap W_0^m(\Omega)$ onto L^2 and, for $r \geq r_0$, suppose that L maps $W_0^{m,r}(\Omega) \cap W^{2m,r}(\Omega)$ continuously onto L^r. Now if $u \in W^{2m}(\Omega) \cap W_0^m(\Omega)$ and $Lu \in L^{\infty}$, then $Lu \in L^r$ for $r < \infty$ and hence $u \in W^{2m,r}(\Omega)$ for $r \geq r_0$ and, by the open mapping theorem,

$$\|u\|_{2m,r} \leq C_r \|Lu\|_{L^r}$$

where $C_r > 0$ does not depend on u. Further, if the boundary of Ω is sufficiently smooth so that the embedding,

$$J : W^{2m,r}(\Omega) \rightarrow C(\Omega),$$

is continuous for $r \geq r_0$, then there exists $E_x \in L^{r'} \subset L^1$,

$1/r + 1/r' = 1$, satisfying,

$$u(x) = \int_\Omega Lu \cdot E_x,$$

for all $x \in \Omega$, and hence E_x is the fundamental solution for the formal adjoint L^* of L. If, for example, L is an elliptic operator with real analytic coefficients, then E_x is real analytic in $\Omega - \{x\}$ ([4.2]). If, in this case, we select distinct points $x_1,\ldots,x_N$ in Ω and let $F_i = E_{x_i}$, $i = 1,\ldots,N$, then $F_1,\ldots,F_N$ satisfy the hypotheses of Theorem 4.1.

Corollary 4.3. Suppose that $W^{2m,r}(\Omega)$ is embedded continuously into $C(\Omega)$ for all r exceeding some r_0 and that L is a continuous bijection mapping $W^{2m}(\Omega) \cap W_0^m(\Omega)$ onto L^2 and that L maps $W^{2m,r}(\Omega) \cap W_0^{m,r}(\Omega)$ continuously onto L^r for $r \geq r_0$. Let $p_1,\ldots,p_N$ be distinct points of Ω with $E_{p_1},\ldots,E_{p_N}$ linearly independent over any set of positive measure. Let $\underline{r} \in R^N$ and let

$$U = \{u \in W^{2m}(\Omega) \cap W_0^m(\Omega) : u(p_i) = r_i,\ i = 1,\ldots,N \text{ and } Lu \in L^\infty\}.$$

If $\psi \in L^\infty$, then there is a unique solution $\varphi \in U$ to the minimization problem

$$\alpha = \inf\{\|Lu-\psi\|_{L^\infty} : u \in U\} \tag{4.2}$$

and $L\varphi - \psi$ has constant modulus a.e. on Ω.

Example 4.1. Let $n = 3$ and let $\Omega = \{x \in R^3 : |x| < 1\}$. Let L be the Laplacian and $U = \{u \in W^2 \cap W_0^1 : u(0) = 1\}$. According to Corollaries 4.2 and 4.3, there is a unique function $\varphi \in U$ with $\|L\varphi\|_{L^\infty}$ minimal over U. In this case, it is not difficult to see that $\varphi(x) = 1 - |x|^2$ and $L\varphi = -6$ on Ω. Note that φ is real analytic in accordance with Corollary 4.2, (iii).

Example 4.2. Let Ω and L be as in Example 1 and let $a = (1/2,0,0)$. Let $U = \{u \in W^2 \cap W_0^1 : u(a) = 1 \text{ and } u(-a) = -1\}$. Suppose φ is the solution of (4.1) for this Ω, L and U with $\psi = 0$. Let $\varphi_1(x) = -\varphi(-x)$. Then $\varphi_1 \in U$ and $\|\Delta\varphi_1\|_{L^\infty} = \|\Delta\varphi\|_{L^\infty}$ and so by uniqueness, $\varphi_1 = \varphi$; thus, $\Delta\varphi(x) = -\Delta\varphi(-x)$. Hence, since $|\Delta\varphi|$ is constant, and not zero, the sets P and N are both non-empty. Furthermore, if we let $\varphi_2(x_1,x_2,x_3) = -\varphi(-x_1,x_2,x_3)$, then $\varphi_2 \in U$ and $|\Delta\varphi_2| = |\Delta\varphi|$. Thus $\varphi_2 = \varphi$ and so P and N are separated by the plane $x_1 = 0$. This clearly implies that φ is not C^∞ on all of Ω and hence shows that the set Ω_1 in Corollary 4.2 can really be a proper subset of Ω.

Example 4.3. The condition that $F_1,\ldots,F_N$ be linearly independent over any set of positive measure is, in fact, a necessary condition in order that the conclusions of Theorem 4.1 hold.

To see this suppose first that $F_N = 0$ on E where E has positive measure. Let $A = \sup\{|\int_\Omega fF_N| : \|f\|_{L^\infty} \leq 1,\ f = 0 \text{ on } E, \text{ and } \int_\Omega fF_i = 0 \text{ for } j = 1,\ldots,N-1\}$. Clearly there is an f_0 which vanishes a.e. on E, $\|f_0\|_{L^\infty} = 1$, $\int_\Omega f_0F_i = 0$ for $j = 1,\ldots,N-1$ and $\int_\Omega f_0F_N = A$. If g vanishes off E, $\|g\|_{L^\infty} \leq 1$, and $\int_\Omega gF_i = 0$ for $j = 1,\ldots,N-1$, then $f_0 + g$ is also bounded by 1 and satisfies $\int_\Omega (f_0+g)F_i = 0$ if $1 \leq j \leq N-1$ and $\int_\Omega (f_0+g)F_N = A$. Let

$$U = \{u \in W^{2m}(\Omega) \cap W_0^m(\Omega) : \int_\Omega LuF_i = 0 \text{ if } 1 \leq j \leq N-1 \text{ and } \int_\Omega LuF_N = A\}$$

and take $\psi = 0$. Then $\inf\{\|Lu\|_{L^\infty} : u \in U\} = 1$ and there are infinitely many $u \in U$ which satisfy $\|Lu\|_{L^\infty} = 1$.

In general, if $\Sigma_{1 \leq i \leq N}\, c_iF_i = 0$ on E where without loss of generality $c_N = 1$, let $G_j = F_j + \Sigma_{1 \leq i \leq N-1}\, c_iF_i$ for $i = 1,\ldots,N$. By the above there is an f_0 supported off E with $\|f_0\|_{L^\infty} = 1$,

$\int_\Omega f_0 G_i = 0$ if $1 \leq i \leq N - 1$ and $\int_\Omega f_0 G_N = \max\{|\int_\Omega f G_N| : \|f\|_{L^\infty} \leq 1,$ $f = 0$ on E, $\int_\Omega f G_i = 0$, $i = 1,\ldots,N - 1\}$. Let $r_i = \int_\Omega f_0 F_i$, $i = 1,\ldots,N$ and set

$$U = \{u \in W^{2m}(\Omega) \cap W_0^m(\Omega) : \int_\Omega LuF_i = r_i,\ i = 1,\ldots,N\}.$$

Then $\inf\{\|Lu\|_{L^\infty} : u \in U\} = 1$ and there are infinitely many $u \in U$ which satisfy $\|Lu\|_{L^\infty} = 1$.

Remarks. This chapter has essentially been extracted from the authors' paper [4.1]. The smoothness of the boundary $\Omega - (\Omega_+ \cup \Omega_-)$ would be of interest in the general case. Also, we note that Example 4.3 shows that the hypothesis of linear independence, on sets of positive measure, of $F_1,\ldots,F_N$ is both a necessary and sufficient condition for Theorem 4.1.

REFERENCES

4.1 S. D. Fisher and J. W. Jerome, "Elliptic variational problems in L^2 and L^∞," Indiana Journal of Math., 23 (1974), 685-698.

4.2 A. Friedman, Partial Differential Equations, Holt, Rinehart and Winston, New York, 1969.

4.3 F. John, "General properties of solutions of linear elliptic partial differential equations," in, Proceedings, Conference on Partial Differential Equations, (N. Aronszajn, ed.), Stillwater, Oklahoma (1951), pp. 113-175.

§5. L^1 Minimization in One and Several Variables

Let $a = x_0 < x_1 < \dots < x_m = b$ be a fixed partition of the closed interval $[a,b]$ and let U be the closed flat in the Sobolev space $W^{n,1}(a,b)$, $1 \leq n \leq m + 1$, defined by the interpolation of specified values r_i at the points x_i, $i = 0,1,\dots,m$. If

$$(5.1) \qquad \alpha = \inf\{\|D^n f\|_{L^1(a,b)} : f \in U\}.$$

then the minimization problem (5.1) does not, in general have an interpolating solution in the class $W^{n,1}(a,b)$. More generally, if U is a closed flat in a Banach space X and R is a continuous linear mapping of X onto Y with finite-dimensional null space then it is possible that $\inf\{\|Ru\|_Y : u \in U\}$ is not attained in U if Y is not reflexive, for example. In his recent paper [5.3], Holmes has discussed, with both a literature survey and new results, the technique of embedding such optimization problems in dual spaces. By considering the extended problem of minimizing $\|R^{**}\varphi\|_{Y^{**}}$ over the flat JU in X^{**}, where J denotes the natural injection of X into X^{**}, Holmes has shown that a solution φ exists in JU, achieving the same norm extremal value as in the original problem under a natural poisedness hypothesis. Thus, the problem has a solution in the sense of Ioffe and Tihomirov [5.4].

In this chapter we shall discuss a concrete way in which such problems can be extended to possess natural solutions. In particular, for the problem (5.1), we expand the class $W^{n,1}(a,b)$ to include functions whose nth. derivatives are measures. The extended problem has a solution s such that Var $D^{n-1}s$ coincides with the extremal value of (5.1). s has the property that $s \in C^{n-2}[a,b]$ and is a spline function of degree $n - 1$. This occurs because the measure $D^n s$ has minimal support, concentrated at $m + 1$ or fewer points, as we shall show. Our results are very general and apply in several variables as well as one variable, with appropriate operators. Also, more generally, U may be a slab defined by arbitrary continuous linear inequality constraints in (5.1) and D^n may be replaced by more general differential operators.

The plan of the chapter is as follows. In section 5.1 we present the basic existence theorem concerning extremal measures in general context. This convex set of extremal measures, all of which have minimal total variation, is shown to satisfy the hypotheses of the Krein-Milman theorem so that there is at least one extremal measure, whose support is shown to be a finite point set. Univariate and multivariate applications of this result are given in section 5.2, 5.3 and 5.4 together with a linear programming application.

5.1 A Theorem on Constrained Extremal Measures With Minimal Support

Let X be a compact metric space, C(X) the Banach space of real-valued continuous functions on X in the supremum norm and M(X) the dual Banach space of real (finite, regular) Baire measures on X in the total variation norm. Let N be a finite dimensional subspace of C(X) and let H be the direct sum of M(X) and N.

<u>Theorem 5.1</u>. Let $L_0,\dots,L_m$ be linearly independent linear functionals on H of the form

$$L_i(\lambda,P) = \int_X F_i d\lambda + \ell_i(P)$$

where F_i is continuous on X, $i = 0,\dots,m$ and $\{\ell_i\}_{i=0}^m$ is a set of linear functionals defined on N such that P = 0 if $P \in N$ and $\ell_i(P) = 0$ for each $i = 0,\dots,m$. Let $I_0,\dots,I_m$ be compact intervals in $\mathbb{R}$, each possibly consisting of a single point, and let

$$U = \{(\lambda,P) \in H : L_i(\lambda,P) \in I_i,\ i = 0,1,\dots,m\}$$

Set

$$(5.2) \qquad \alpha = \inf\{\|\lambda\| : (\lambda,P) \in U\}.$$

Then U contains at least one pair (λ,P) for which $\|\lambda\| = \alpha$; the set S of such pairs is convex and compact in the weak-$*$ topology. The extreme points of S are all of the form $(\Sigma_0^r c_j\delta_j, P)$ where $r \leq m$, δ_j is the unit point mass at $t_j \in X$ for $j = 0,\ldots,r$ and $\Sigma_0^r |c_j| = \alpha$.

Proof. Let $\{g_\nu = (\lambda_\nu, P_\nu)\}$ be a sequence in U with $\|\lambda_\nu\| \to \alpha$. Since $L_i(g_\nu) \in I_i$ for all ν and since $\int_X F_i d\lambda_\nu$ is also uniformly bounded for all ν and i, we find that

$$(5.3) \qquad |\ell_i(P_\nu)| \leq C, \ i = 0,\ldots,m \text{ and } \nu = 1,2,\ldots$$

Hence, by the completeness of $\{\ell_i\}_0^m$ it follows from (5.3) that the sequence P_ν is bounded in norm. Thus, there is a subsequence of $\{P_\nu\}$, denoted $\{P_{\nu_j}\}$, and a $P_0 \in N$ with $\ell_i(P_{\nu_j}) \to \ell_i(P_0)$ as $j \to \infty$ for $i = 0,\ldots,m$. Likewise, the measures $\{\lambda_{\nu_j}\}$ have a weak* accumulation point λ_0 with $\|\lambda_0\| \leq \alpha$. It is easy to check that $(\lambda_0, P_0) \in U$ and hence $\|\lambda_0\| = \alpha$. Thus we have shown that S is nonempty. The convexity of S follows from the convexity of U and the definition of α. Now the convex set T of measures determined as the set of first components of S is clearly weak* closed in M(X) since it is bounded and contains all its weak* sequential limit points. (Since X is compact metric, C(X) is separable [5.2; Theorem 14-9.15, p. 276] and hence the closed unit ball of M(X) is metrizable in the weak-$*$ topology [2.4; p. 426]). Now by the Krein-Milman theorem let (λ,P) be any extreme point of S and suppose there are $m + 2$ disjoint Baire sets $E_0,\ldots,E_{m+1}$ in X which have positive λ-measure. Let λ_i be the restriction of λ to E_i for $i = 0,\ldots,m + 1$ and let v_i be the vector in $(m+1)$-space whose jth. coordinate is $L_j(\lambda_i, P)$, $j = 0,\ldots,m$ and

$i = 0,\ldots,m+1$. The vectors $v_0,\ldots,v_{m+1}$ must be linearly dependent in R^{m+1} and hence there are scalars $a_0,\ldots,a_{m+1}$, not all zero, with $\Sigma_0^{m+1} a_i v_i = 0$. Let $\mu = \Sigma_0^{m+1} a_i\lambda_i$ so that μ is not the zero measure. Then we have, for $0 \neq \epsilon$ sufficiently small,

$$\|\lambda+\epsilon\mu\| = \|\lambda\|_{X-\cup_j E_j} + \Sigma_0^{m+1}(1+\epsilon a_i)\|\lambda_i\| = \alpha + \epsilon\, \Sigma_0^{m+1} a_i\|\lambda_i\|$$

Now, if $\Sigma_0^{m+1} a_i\|\lambda_i\| \neq 0$, then some choice of ϵ gives a pair $(\lambda+\epsilon\mu, P)$ in U with $\|\lambda+\epsilon\mu\| < \alpha$, a contradiction. Hence, $\Sigma_0^{m+1} a_i\|\lambda_i\| = 0$ so that each of $(\lambda+\epsilon\mu, P)$, $(\lambda-\epsilon\mu, P)$ lies in S. The convex representation

$$(\lambda, P) = \tfrac{1}{2}(\lambda-\epsilon\mu, P) + \tfrac{1}{2}(\lambda+\epsilon\mu, P)$$

then contradicts the choice of (λ, P) as an extreme point of S. It follows that there are at most $m + 1$ disjoint subsets of X of positive λ measure and the theorem follows. $\square$

5.2 Univariate Generalized Splines

Let $I = [a,b]$ be a closed interval in $\mathbb{R}$ and let L be a nonsingular linear differential operator of order n on I of the form

$$L = D^n + \sum_{j=0}^{n-1} a_j D^j, \quad n \geq 2,$$

where $a_j \in C(I)$, $j = 0,\ldots,n-1$. If $f \in W^{n,1}(I)$ then the representation

$$(5.4) \qquad f(x) = P(x) + \int_a^b \hat{\theta}(x,\xi)\, Lf(\xi)\, d\xi, \quad a \leq x \leq b$$

holds, where P in the null space N_L of L is defined by $D^jP(a) = D^jf(a)$, $0 \leq j \leq n-1$ and where the function $\theta(\cdot,\xi) \in N_L$ is defined for each $\xi \in [a,b]$ by

$[D_x^j \theta(x,\xi)]_{x=\xi} = \delta_{j,n-1}$, $0 \leq j \leq n-1$, and $\hat{\theta}$ is given by

$$\hat{\theta}(x,\xi) = \begin{cases} \theta(x,\xi) & \text{if } a \leq \xi \leq x \leq b \\ 0 & \text{otherwise} \end{cases}$$

Here $W^{n,1}(I)$ is the real Sobolev class of $f \in C^{n-1}(I)$ such that $D^{n-1}f$ is absolutely continuous and $D^n f \in L^1(I)$. In this application we shall have need of the larger class $Q^n(I) = \{f : Lf \in M(I)\}$ where Lf is taken in the weak sense and M(I) is the space of real Baire measures on I. Equivalently,

$$(5.5) \qquad f(x) = P(x) + \int_a^b \hat{\theta}(x,\xi)\, d\mu(\xi)$$

for $\mu \in M(I)$ and $P \in N_L$ if $f \in Q^n(I)$. We remark that

$$\|Lf\|_{L^1(I)} = \|Lf\|_{M(I)} \text{ if } f \in W^{n,1}(I).$$

Now let $\Lambda = \{\lambda_0,\ldots,\lambda_m\}$ be a linearly independent set of linear functionals on $C^{n-2}(I)$ for which the Peano representations

$$(5.6) \qquad \lambda f = \lambda P + \int_a^b \lambda\hat{\theta}(\cdot,\xi)\, Lf(\xi)\, d\xi$$

hold for every $\lambda \in \Lambda$ and $f \in W^{n,1}(I)$ with $\{\lambda\hat{\theta}(\cdot,\xi)\}_{\lambda \in \Lambda}$ a linearly independent set of continuous functions in ξ on I where P is given by (5.4). Since $L^1(I)$ is weak* dense in M(I), (5.6) holds over the larger class $Q^n(I)$. Let $I_0,\ldots,I_m$ be compact intervals in $\mathbb{R}$, which are possibly single points, and let

$$U = \{f \in W^{n,1}(I) : \lambda_i(f) \in I_i,\ i = 0,\ldots,m\}.$$

Theorem 5.2. Let $\alpha = \inf\{\|Lf\|_{L^1(I)} : f \in U\}$. Then there are distinct points $t_0 < \ldots < t_r$ in I, $r \leq m$, real numbers $a_0,\ldots,a_r$ and a function $P \in N_L$ such that the generalized spline function

$$s(x) = \sum_{j=0}^{r} a_j \hat{\theta}(x,t_j) + P(x)$$

satisfies $\lambda_i s \in I_i$, $i = 0,\ldots,m$ and $\sum_{j=0}^{r} |a_j| = \alpha$, i.e., $\|Ls\|_{M(I)} = \alpha$. s solves the extended minimization problem $\alpha = \|Ls\|_{M(I)} = \inf\{\|Lf\|_{M(I)} : f \in \tilde{U}\}$. Here $\tilde{U} \supset U$ consists of all $f \in Q^n(I)$ satisfying $\lambda_i f \in I_i$, $i = 0,\ldots,m$. Finally,

$$Ls = \sum_j a_j \delta(\cdot,t_j) \tag{5.7}$$

where $\delta(\cdot,t)$ represents the Dirac delta functional at t.

Proof. $W^{n,1}(I)$ is algebraically isomorphic to $L^1(I) + N_L$ under (5.4). Now $L^1(I) + N_L$ is weak* dense in $M(I) + N_L$ so that

$$\tilde{\alpha} = \inf\{\|Lf\|_{M(I)} : f \in \tilde{U}\} = \alpha$$

Indeed, if $\mu_0 = Lf_0$ is chosen so that $\|\mu_0\| = \tilde{\alpha}$, $f_0 \in \tilde{U}$, choose $\psi_\nu \xrightarrow{*} \mu_0$ such that $\psi_\nu \in L^1(I)$, $\nu = 1,2,\ldots$ and $\|\psi_\nu\|_{L^1(I)} \leq \|\mu_0\|_{M(I)}$. We show how to adjust ψ_ν slightly so as to obtain an element which, upon integration, is in U.

Thus, we consider the map $K : L^1(I) \to \mathbb{R}^{m+1}$ given by

$$Kg = \left(\int_I F_0 g\,dx,\ldots, \int_I F_m g\,dx\right)$$

where $F_i(\xi) = \lambda_i \hat{\theta}(\cdot,\xi)$, $i = 0,\ldots,m$. K is clearly onto by the linear independence of the F_i and the quotient space $L^1(I)/\ker K$ is algebraically and topologically isomorphic to $\mathbb{R}^{m+1}$. Thus, there exist sequences $\epsilon_\nu \to 0$ and $\{\varphi_\nu\} \subset L^1(I)$ such that $\|\varphi_\nu\| \leq \epsilon_\nu$ and $\varphi_\nu + \psi_\nu \in LU$, $\nu = 1,2, \ldots$. Since

$$\alpha \leq \liminf_{\nu \to \infty} \|\psi_\nu + \varphi_\nu\|_{L^1(I)} = \liminf_{\nu \to \infty} \|\psi_\nu\|_{L^1(I)} \leq \|\mu_0\|_{M(I)} = \tilde{\alpha}$$

it follows that $\alpha = \tilde{\alpha}$. The result now follows from Theorem 1.1 and the representation (5.5). □

Corollary 5.3. Consider the extremal problem (5.1) of the introduction. Then there is a polynomial spline function of degree $n - 1$ in $C^{n-2}[a,b]$ with at most $m + 1$ knots in (a,b) satisfying $\mathrm{Var}\, D^{n-1}s = \alpha$ and $s(x_i) = r_i$, $i = 0,\ldots,m$.

5.3 Multivariate Generalized Spline Functions

Let Ω be a bounded domain in $\mathbb{R}^\ell$, $\ell \geq 2$, and let L be a linear differential operator of order n. Our fundamental assumption concerning L is that the mapping

$$L : W^{n,1}(\Omega) \to L^1(\Omega)$$

is continuous and surjective. Here $W^{n,1}(\Omega)$ is the Sobolev space of functions f with distribution derivatives $D^{\alpha'}f \in L^1(\Omega)$, $|\alpha'| \leq n$, with norm

$$\|f\|_{W^{n,1}(\Omega)} = \sum_{|\alpha'| \leq n} \|D^{\alpha'}f\|_{L^1(\Omega)} \tag{5.8}$$

We shall further assume that there is a closed linear subspace F of $W^{n,1}(\Omega)$ such that the restriction of L to F admits a unique inverse representation of the form

$$f(x) = \int_{\Omega} G(x,\xi) Lf(\xi) d\xi, \quad f \in F, \tag{5.9}$$

where we explicitly assume that $G(x,\cdot) \in C(\bar{\Omega})$ for each $x \in \Omega$.

We have in mind, of course, the specific application where L is a uniformly elliptic operator of even order $n = 2k$, in which case, if $n > \ell$, the above hypotheses are satisfied for sufficiently smooth boundary $\partial\Omega$, for the choice $F = W^{2k,1}(\Omega) \cap W_0^{k,1}(\Omega)$ and $L + \lambda I$, if λ is sufficiently large. Here $W_0^{k,1}(\Omega)$ is the completion, in the norm (5.8), of the $C^{\infty}(\Omega)$ functions with compact support in Ω.

Let $\Lambda = \{\lambda_0,\ldots,\lambda_m\}$ be any linearly independent set of linear functionals on F such that the Peano representations

$$\lambda f = \int_{\Omega} \lambda G(\cdot,\xi) Lf(\xi) d\xi, \quad f \in F,$$

hold with $\lambda G(\cdot,\xi) \in C(\bar{\Omega})$. Then, if $I_0,\ldots,I_m$ are compact intervals in R and

$$U = \{f \in F : \lambda_i f \in I_i, \; i = 0,\ldots,m\},$$

we may state the multivariate analogue of the previous theorem.

<u>Theorem 5.4</u>. Let $\alpha = \inf\{\|Lf\|_{L^1(\Omega)} : f \in U\}$. Then there are distinct points $t_0,\ldots,t_r$ in $\bar{\Omega}$, $r \leq m$, and real numbers $a_0,\ldots,a_r$ such that the multivariate generalized spline function

$$s(x) = \sum_{j=0}^{r} a_j G(x,t_j)$$

satisfies $\lambda_i s \in I_i$, $i = 0,\ldots,m$ and $\sum_{j=0}^{r} |a_j| = \alpha$, i.e., $\|Ls\|_{M(\Omega)} = \alpha$. s solves an extended minimization problem as before and satisfies the relation (2.4).

5.4 A Mathematical Programming Application

Discrete spline functions were introduced by Mangasarian and Schumaker [5.5] as minimizing a general forward difference operator $L : \mathbb{R}^{\ell} \to \mathbb{R}^{\ell-n+1}$ subject to affine constraints in the ℓ^p norm, $1 \leq p \leq \infty$. Methods of mathematical programming were employed in the existence theory to deduce the closure of certain sets. We shall show here that, in the special case $p = 1$, there is a solution s such that Ls has support confined to a subspace of $\mathbb{R}^{\ell}$ of dimension $m + 1$ if there are $m + 1$ constraint functionals.

Specifically, let L be of the form

$$(5.10) \qquad (Lx)_j = \sum_{\nu=1}^{n} a_\nu x_{\nu+j-1}, \quad j = 1,\ldots,\ell - n + 1$$

where $a_n \neq 0$. Then L maps $\mathbb{R}^{\ell}$ onto $\mathbb{R}^{\ell-n+1}$ and, by the rank-nullity theorem, the null space N of L is of dimension $n - 1$. The complement M of N in R^{ℓ} is of dimension $\ell - n + 1$ and L maps M bijectively onto $R^{\ell-n+1}$. Thus, if $\Lambda = \{\lambda_0,\ldots,\lambda_m\}$ is a linearly independent set of linear functionals on $\mathbb{R}^{\ell}$ of the form

$$(5.11) \qquad \lambda_i x = \sum_{j=1}^{\ell} a_{ij} x_j, \quad x \in \mathbb{R}^{\ell},$$

$i = 0,\ldots,m$, such that $N \subset \operatorname{span}_{0 \leq i \leq m} \{a_{i1},\ldots,a_{i\ell}\}$ and if $I_0,\ldots,I_m$ are compact intervals in $\mathbb{R}$ and

$$U = \{x \in \mathbb{R}^{\ell} : \lambda_i x \in I_i, \; i = 0,\ldots,m\}$$

then we have the following consequence of Theorem 5.1.

<u>Theorem 5.5</u>. Let $\alpha = \inf\{\sum_{j=1}^{\ell-n+1} |(Lx)_j| : x = (x_1,\ldots,x_\ell) \in U\}$. Then there exists a vector $s \in U$ satisfying $\sum_{j=1}^{\ell-n+1} |(Ls)_j| = \alpha$ and,

moreover, the support of Ls is confined to at most m + 1 components.

Proof. Take X to be ℓ - n + 1 distinct points with the discrete topology. □

Remarks. This chapter is excerpted from the authors' paper [5.1]. The point made here is simply that enlargement of the admissible set leads to solutions without decreasing α; moreover, the extreme points of the solution set are integral transforms of minimal support measures. The results are valid in one and several variables and even more generally, as Theorem 5.1 shows, and hold for inequality as well as equality constraints for arbitrary continuous linear functionals. For equality constraints, the problem may be reduced by duality theory to the problem of norm preserving extension of continuous linear functionals. In this context, a theorem of Singer [5.6, Ch. 2] applies to yield a result structurally like the present one.

REFERENCES

5.1 S. D. Fisher and J. W. Jerome, "Spline solutions to L^1 extremal problems in one and several variables," J. Approximation Theory, 13 (1975), 73-83.

5.2 A. Gleason, Fundamentals of Abstract Analysis, Addison-Wesley, Reading, Mass., 1966.

5.3 R. B. Holmes, "R-splines in Banach spaces: I," J. Math. Anal. Appl., 40 (1972), 574-593.

5.4 A. Ioffe and V. Tihomirov, "Extension of variational problems," Trans. Moscow Math. Soc., 18 (1968), 207-273.

5.5 O. L. Mangasarian and L. L. Schumaker, "Discrete splines via mathematical programming," SIAM J. Control, 9 (1971), 174-183.

5.6 I. Singer, "Dea Mai Buna Approximare in Spatii Vectoriale Normate prin Elementa din Subspatii Vectoriale," Editura Acad. Republ. Social, Romania, Bucuresti, 1967.

PART III. UNIQUENESS THEOREMS

§6. Sets of Uniqueness in L^∞ Minimization Problems

6.1 A General Theorem

Theorem 6.1. Let X be the dual of a separable Banach space Y and let T be a linear weak-* continuous mapping of X into $L^\infty(\Omega,\mu)$ with closed range and finite dimensional null space. Let $L_1,\ldots,L_m$ be elements of Y, let K be a compact convex subset of $\mathbb{R}^m$ and define

$$U = \{x \in X : (L_1(x),\ldots,L_m(x) \in K\}$$

Then the extremal problem

$$\alpha = \inf\{\|Tx\|_\infty : x \in U\} \tag{6.1}$$

has a solution. Further, there is a set E of positive measure in Ω such that $Tx = Ty$ a.e. μ on E for any solutions x, y of (6.1).

Proof. The range of T is weak-* closed since T is weak-* continuous and has closed range. Also, the existence of solutions follows from Theorem 2.5. If we put $U_0 = \{y \in X : L_j(y) = 0 \text{ for } 1 \le j \le m\}$, then, as in the proof of Corollary 3.4, TU_0 is a weak-* closed subspace of L^∞ so that by Theorem 3.3, there is a set $E = E(x_0)$ of positive measure on which $|Tx_0| = \alpha$ a.e. for each solution x_0. Let S be the set of all solutions to (6.1). S is convex and hence so is TS; if we topologize TS by the weak-* topology, then TS is also compact. Since the unit ball of L^∞ in the weak-* topology is metrizable [2.4, p. 426], TS is a compact metric space and hence separable. Let $\{Tx_j\}$ be a countable dense set in TS. Let

$$\hat{x} = \Sigma_1^\infty 2^{-j} x_j$$

Then $\hat{x} \in S$ and hence there is a set E of positive measure on which $|T\hat{x}| = \alpha$ a.e. This immediately implies that $Tx_j = Tx_i$ on E for all i, j and, of course, $|Tx_i| = \alpha$ a.e. on E. Since $\{Tx_j\}$ is weak-* dense in TS, it follows that $Tx = Tx_i$ on E for any solution x. $\square$

6.2 Applications

Example 6.1. Integration Functionals

Let $I_1,\ldots,I_m$, $m \geq n$, be disjoint closed sets of positive measure in $[a,b]$; let $L = D^n$, $X = W^{n,\infty}(a,b)$ and define

$$L_j f = \int_{I_j} Lf(t)\,dt, \quad j = 1,\ldots,m.$$

Then $L_1,\ldots,L_m$ are linearly independent elements of the dual space of $W^{n,\infty}(a,b)$. The set E asserted in Theorem 6.1 satisfies $I_{j_0} \subset E \subset \bigcup_{j=1}^{m} I_j$ for some $1 \leq j_0 \leq m$. To see this, recall from Theorem 3.3 that for any solution f_0 to (6.1) there is a function $g \in L^1(a,b)$ with

$$0 = \int_a^b gLv$$

for each $v \in U_0$ and $gf_0^{(n)} \geq 0$, $|f_0^{(n)}| = \alpha$ where $g \neq 0$. Such a g must necessarily satisy the equation

$$L^a g = \sum_1^m \beta_j L_j$$

where L^a is the operator adjoint of $L = D^n$ and $\beta_1,\ldots,\beta_m$ are scalars. Hence, for $f \in W^{n,\infty}(a,b)$,

$$\int_a^b (Lf)g = \int_a^b L^a g(f) = \sum_1^m \beta_j L_j(f) = \int_a^b (Lf)(\sum_1^m \beta_j \chi_j)$$

where χ_j is the characteristic function of I_j. Hence $g = \sum_1^m \beta_j \chi_j$ so that g vanishes off $\bigcup_1^m I_j$ and $g = \beta_j$ on I_j, $j = 1,\ldots,m$. Hence, $|f_0^{(n)}| = \alpha$ on at least one I_j since $\beta_j \neq 0$ for some j, $1 \leq j \leq m$. If K is the single point $\{|I_1|,\ldots,|I_m|\}$, then clearly the solution to (6.1) is $f^{(n)} \equiv 1$ on $I_1 \ldots I_m$ and hence in this case $E = \bigcup_1^m I_j$.

In order to obtain more precise information on the solutions of L^∞ extremal problems we must make more assumptions.

<u>Corollary 6.2</u>. Let $X = W^{n,\infty}(a,b)$, let $L = D^n + \sum_{j=0}^{n-1} a_j D^j$ where $a_j \in C^j[a,b]$, and suppose that the linear functionals $L_1,\ldots,L_m$ of Theorem 6.1 are given by linear combinations of derivatives as in (3.13) of order n - 1 or less at a fixed set of points $x_1,\ldots,x_r$, $a \leq x_1 < x_2 < \ldots < x_r \leq b$. Then there is a solution f of (6.1) and there is at least one interval (x_i,x_{i+1}) on which Lf = Lg for any other solution g of (6.1).

Proof. Let f be a solution of (6.1) guaranteed by Theorem 6.1. We know by Theorem 3.3 that there is a function $g \in L^1$ with $\|g\| = 1$, $0 = \int_a^b gLv$ for all $v \in U_0$, and $|Lf| = \alpha$ where $g \neq 0$. The assumptions on $L_1,\ldots,L_m$ and the coefficients of L imply that U_0 contains all C^∞ functions with compact support in $[a,b] - \{x_i\}_{i=1}^r$ and that L*, the formal adjoint of L, is a well-defined differential operator. Hence,

$$0 = \int_a^b gLv$$

for all $v \in C_0^\infty([a,b]-\{x_i\})$ so that g satisfies $L^*g = 0$ on each interval (x_j,x_{j+1}), $j = 1,\ldots,r-1$. Thus either $g \equiv 0$ on (x_j,x_{j+1}) or g has only a discrete set of zeros in (x_j,x_{j+1}). Since $\|g\| = 1$, g can not vanish on all the intervals and hence there is at least one interval (x_i,x_{i+1}) on which $|Lf| = \alpha$ a.e. The conclusion of the corollary now follows immediately as in the proof of Theorem 6.1. □

6.3 The Fundamental Interval of Uniqueness for Extended-Hermite-Birkhoff Interpolation

Definition. Let points $x_1,\ldots,x_r$ be given, $a \leq x_1 < x_2 < \ldots < x_r \leq b$. Let $1 \leq k_i \leq n$ be given integers, $i = 1,\ldots,r$, and define

$$L_{ij}(h) = \sum_{\nu=0}^{n-1} a_{ij}^{(\nu)} h^{(\nu)}(x_i)$$

for $j = 1,\ldots,k_i$ and $i = 1,\ldots,r$ where the k_i n-tuples $(a_{ij}^{(0)},\ldots,a_{ij}^{(n-1)})$ are linearly independent. We define n_0 to be the maximum positive integer satisfying the following property: for any n_0 consecutive points among $x_1,\ldots,x_r$ the sum of the integers k_i associated with these points does not exceed n. We define the functionals $\{L_{ij}\}$ to be consistent with respect to the given differential operator of order n if the following two conditions hold:

(a) $\sum_{i=1}^{r} k_i \geq n + 1$

(b) for every m consecutive points $x_{s+1},\ldots,x_{s+m}$ and prescribed values y_{ij} there is a function u in the null-space of L satisfying $L_{ij}(u) = y_{ij}$ for $j = 1,\ldots,k_i$ and $i = s + 1,\ldots,s + m$ if $\sum_{i=s+1}^{s+m} k_i \leq n$.

We define the functionals L_{ij} to be completely consistent if they are consistent and, in addition, whenever $x_s,\ldots,x_{s+n_0}$ are $n_0 + 1$ points for which $\sum_{i=s}^{s+n_0} k_i \geq n + 1$, then the only u in the null space of L which satisfies $L_{ij}(u) = 0$ for $j = 1,\ldots,k_i$ and $i = s,\ldots,s + n_0$, is $u \equiv 0$.

Theorem 6.3. Let the functionals L_{ij} be consistent and let L be the differential operator of Corollary 6.2. Then there is an interval J of the form (x_k, x_ℓ) where $\ell - k \geq n_0$ on which $Lf = Lg$ for any solution f, g of (6.1) and $|Lf| = \alpha$; further, $\sum_{i=k}^{\ell} k_i \geq n + 1$. If the functionals are completely consistent, then $f = g$ on J.

The proof of Theorem 6.3 requires the following lemma which isolates the key construction.

Lemma 6.4. Suppose that the given functionals L_{ij} are consistent. Let $\ell \geq 1$, let $x_{\ell+1}, \ldots, x_{\ell+m}$ be m consecutive nodes and suppose $\sum_{i=\ell+1}^{\ell+m} k_i = s \leq n$. If $\underline{r} \in \mathbb{R}^s$ and $E \subset (x_\ell, x_{\ell+1})$ has positive measure, then there is an $h \in W^{n,\infty}(a,b)$ with $h \equiv 0$ on $[a, x_\ell]$, Lh is supported in E, and $L_{ij}h = r_{ij}$, $j = 1, \ldots, k_i$, and $i = \ell + 1, \ldots, \ell + m$.

Proof. Define the kernel $\theta(\cdot,\cdot)$ on $[a,b] \times [a,b]$ as the unique function $\theta(\cdot,\xi) \in N_L$, for each fixed ξ, satisfying

$$D^j\theta(\cdot,\xi)]_\xi = \delta_{j,n-1}, j = 0, \ldots, n-1$$

and set

$$\hat{\theta}(x,\xi) = \begin{cases} \theta(x,\xi) & \text{if } x \geq \xi \\ 0 & \text{otherwise.} \end{cases}$$

Then the representation

$$f(x) = p(x) + \int_a^b \hat{\theta}(x,\xi) Lf(\xi)\, d\xi$$

holds for $x \in [a,b]$ where $p \in N_L$ is determined from $D^\nu p(a) = D^\nu f(a)$, $\nu = 0, \ldots, n-1$. Now the formal adjoint L^* of L exists and [6.2, pp. 75-78]

$$\text{(6.2)} \qquad \theta(x,\xi) = \sum_{j=1}^{n} \varphi_j(x)\varphi_j^*(\xi)$$

for some choice $\{\varphi_1,\ldots,\varphi_n\}$ of a basis of the null space N_L of L and some choice $\{\varphi^*,\ldots,\varphi^*_n\}$ of a basis of the null space of L*.

If $g \in L^\infty(E)$ the function,

$$(6.3) \qquad f(x) = \int_{x_\ell}^{x} \theta(x,y)g(y)\,dy, \quad x \geq x_\ell,$$

gives the unique function in $W^{n,\infty}(a,b)$ which vanishes for $x \leq x_\ell$ and for which $Lf = g$. If we substitute the expression (6.2) for θ into (6.3) we find that

$$(6.4) \qquad f(x) = \sum_{j=1}^{n} \varphi_j(x) \int_E g(y)\varphi^*_j(y)\,dy, \quad x \geq x_{\ell+1}.$$

We wish to show that the coefficients of $\varphi_1,\ldots,\varphi_n$ in (6.4) may be any n-tuple of real numbers for an appropriate choice of $g \in L^\infty(E)$. Hence, consider the map T defined by

$$(6.5) \qquad Tg = \{\int_E g(y)\varphi^*_j(y)\,dy\}_{j=1}^{n}, \quad g \in L^\infty(E).$$

T is clearly linear; if T were not onto R^n, then there would be scalars $\beta_1,\ldots,\beta_n$, not all zero, with

$$0 = \sum_{j=1}^{n} \beta_j \int_E g(y)\varphi^*_j(y)\,dy = \int_E g(y)(\sum_1^n \beta_j\varphi^*_j(y))\,dy, \text{ for all } g \in L^\infty(E).$$

This implies that $\Sigma_1^n \beta_j\varphi^*_j$ vanishes a.e. on E and since E has positive measure we learn $\Sigma_1^n \beta_j\varphi^*_j = 0$ which in turn implies all the β_j are zero. Hence, T is onto and we deduce the Lemma from the consistency hypothesis. □

A repeated application of Lemma 6.4 produces the following result which we simply state.

Lemma 6.5. Let n_0 be given as above and suppose the L_{ij} are consistent. Let E be a closed set in $[x_1, x_r]$ with the property that E intersects in a set of positive measure any collection of n_0 consecutive intervals in the collection $\{[x_i, x_{i+1}]\}_{i=1}^{r-1}$. Then, given data at the points $x_1, \ldots, x_r$ there is a function $f \in W^{n,\infty}(x_1, x_r)$ which interpolates those data such that Lf is supported in E.

We turn now to the proof of Theorem 6.3.

Proof of Theorem 6.3. Let f be any solution of (1.2) with $\alpha > 0$ and for $\delta > 0$ let E be the set where $|Lf| \leq \alpha - \delta$. Suppose that E intersects in a set of positive measure any collection of n_0 consecutive intervals in the collection $\{[x_i, x_{i+1}]\}_{i=1}^{r-1}$. By Lemma 6.5, then, there is a function $g \in W^{n,\infty}(a,b)$ with Lg supported on E such that g interpolates the same data as f. Consider for $\varepsilon > 0$ the function $h = (1+\varepsilon)^{-1}(f+\varepsilon g)$. Clearly $h \in W^{n,\infty}(a,b)$ and h lies in U; further, when ε is sufficiently small, $\|Lh\|$ is strictly less than α, a contradiction. It follows that there is a collection of n_0 consecutive intervals on which $|Lf| = \alpha$ a.e. for this solution f.

Let $I_1, \ldots, I_p$ be those intervals among $\{[x_i, x_{i+1}]\}_{i=1}^{r-1}$ with the property that for each I_j, $j = 1, \ldots, p$, there is some solution f_j to (6.1) with $|Lf|_j < \alpha$ on a set $E_j \subseteq I_j$ of positive measure. If this set of intervals is empty, then $J = [x_1, x_r]$. Thus, assume $p \geq 1$. Now let $g = p^{-1}(f_1 + \ldots + f_p)$; then g is a solution to (6.1) and by construction $|Lg| < \alpha$ on a set of positive measure in each of the intervals I_j, $j = 1, \ldots, p$. However, we have already proved that there are n_0 consecutive intervals on which $|Lg| = \alpha$ a.e. Hence, the collection $\{I_j\}_{j=1}^{p}$ omits n_0 consecutive intervals in $[x_1, x_r]$ and hence $|Lf| = \alpha$ a.e. on these n_0 consecutive intervals for any solution f to (6.1).

Let J_1 be the union of the n_0 intervals found above; let J be the largest interval of the form $[x_k, x_{k+\nu}]$ which contains J_1 and such that $|Lf| = \alpha$ a.e. on J for all solutions f of (6.1); it is possible that $J = J_1$. We claim there are $n + 1$ or more of the functionals L_{ij} associated with the points $x_k, \ldots, x_{k+\nu}$. Suppose this is false; we shall construct a solution f of (6.1) with $|Lf| \leq \alpha' < \alpha$ a.e. on J, a contradiction. Let us suppose for simplicity that $k > 1$; minor modifications of the following take care of the case $k = 1$. Suppose also that $k + \nu < r$; let $I_1 = (x_{k-1}, x_k)$ and $I_2 = (x_{k+\nu}, x_{k+\nu+1})$. By the definition of I_1, there is a solution f_1 of (6.1) with $|Lf_1| \leq \alpha - \delta$ on a closed set E_1 of positive measure in I_1 and a solution f_2 of (6.1) with $|Lf_2| \leq \alpha - \delta$ on a closed set E_2 of positive measure in I_2; let $g = 1/2(f_1+f_2)$ so that g is a solution of (6.1) and $|Lg| \leq \alpha - \delta$ (some $\delta > 0$) on both E_1 and E_2. Let h be the element of $W^{n,\infty}(a,b)$ which vanishes identically on $[x_1, x_{k-1}]$ and for which Lh is Lg on J and zero elsewhere. Since the functionals L_{ij} are consistent and since there are n or fewer of them associated with the points $x_k, \ldots, x_{k+\nu}$ there is an $h_1 \in W^{n,\infty}(a,b)$ which vanishes identically on $[x_1, x_{k-1}]$, which interpolates h at $x_k, \ldots, x_{k+\nu}$, and for which Lh_1 is supported on E_1. (This is just the content of Lemma 6.4). Further, there is an $h_2 \in W^{n,\infty}(a,b)$ which vanishes identically on $[x_1, x_{k+\nu}]$, which interpolates $h - h_1$ at $x_{k+\nu+1}, \ldots, x_r$, and for which Lh_2 is supported on E_2, since $h - h_1$ agrees with a member of N_L on $[x_{k+\nu+1}, x_r]$. Consider now the function $f = g - \epsilon(h-h_1-h_2)$; $f \in W^{n,\infty}(a,b)$ and, by construction, $f \in U$. Further,

$$Lf = \begin{cases} (Lg)(1-\epsilon) & \text{on } J, \\ Lg + \epsilon Lh_1 & \text{on } E_1, \\ Lg + \epsilon Lh_2 & \text{on } E_2, \\ Lg & \text{elsewhere.} \end{cases}$$

Hence, for sufficiently small ϵ, Lf is a solution to (6.1); but $|Lf| \leq \alpha' < \alpha$ a.e. on J, a contradiction. Consequently, we learn that there are $n + 1$ or more of the functionals L_{ij} associated with the points $x_k, \ldots, x_{k+\nu}$. Now let the L_{ij} be completely consistent and let f and g be any two solutions to (6.1). Then $Lf = Lg$ a.e. on J by convexity; hence $f - g = \varphi$ on J where φ is in the null space of L. However, $L_{ij}(f) = L_{ij}(g)$ for all the functionals L_{ij} and thus $L_{ij}(\varphi) = 0$ for those $n + 1$ or more functionals associated with the points $x_k, \ldots, x_{k+\nu}$. This implies $\varphi = 0$ by the complete consistency of the functionals and hence $f = g$ on J, as desired. $\square$

Remarks. Theorem 6.1 shows the existence of a fundamental set $E \subset \Omega$ on which any solutions f and g satisfy $Tf = Tg$. (In §4, $E = \Omega$ held). In the applications of section 6.2 determined by a grid, the set E can be represented in terms of the subintervals determined by the grid. In section 6.3, it is shown that, not only $Lf = Lg$ for any two solutions, but, in fact, $f = g$ on the fundamental interval. This fact, which is the substantive content of Theorem 6.3, holds for L^∞ minimization problems, subject to arbitrary convex constraints determined by completely consistent extended-Hermite-Birkhoff linear functionals. The theorem as stated here represents a semantic improvement over the statement of the corresponding results in the authors' paper [6.1], in that it clearly describes the hypotheses required for the fundamental interval of uniqueness (the more extensive results of [6.1] required an additional hypothesis on L^*). We emphasize the generality of the constraints, which need not be equality constraints, for the validity of the result.

REFERENCES

6.1 S. D. Fisher and J. W. Jerome, "The existence, characterization and essential uniqueness of solutions of L^∞ extremal problems," Trans. Amer. Math. Soc., 187 (1974), 391-404.

6.2 E. Kamke, Differential gleichungen. Lösungsmethoden und Lösungen, Teil. 1: Gewöhnliche Differentialgleichungen, 3rd. ed., Geestand Portig, Leipzig, 1944.

PART IV. BANG-BANG CHARACTERIZATION THEOREMS

§7. Bang-Bang Optimal Controls

7.1 Linear Systems with Two-Point Boundary Conditions

In Example 1.5 we presented a nonlinear optimal control problem in the setting of L^p, $1 < p < \infty$, and demonstrated the existence of an optimal control. In this chapter, we shall consider control problems governed by linear systems and we shall determine bang-bang optimal controls as solutions of L^∞ minimization problems. The qualitative behavior of these controls will follow the pattern of §4.

Consider the system

$$(7.1) \qquad X'(t) = A(t)X(t) + Bu(t) \qquad a \leq t \leq b$$

where $A(t)$ is an $n \times n$ matrix whose entries $a_{ij}(t)$ are continuous functions, $X(t)$ is the column vector $\mathrm{col}(x_1(t),\ldots,x_n(t))$, B is a (constant) nonzero column vector and $u \in L^\infty(a,b)$. We impose the boundary conditions

$$(7.2) \qquad \begin{aligned} X(a) &= X_0 \\ X(b) &= X_1 \end{aligned}$$

and seek to find and determine the properties of a function u of the smallest L^∞ norm for which (7.1)-(7.2) has a solution; that is,

$$(7.3) \quad \alpha_0 = \inf\{\|u\|_\infty : \text{(7.1) is solvable with boundary conditions (7.2)}\}.$$

Clearly, the associated X will have absolutely continuous components with essentially bounded derivatives. When

$$A(t) = \begin{bmatrix} 0 & 1 & 0 & \dots & 0 \\ 0 & 0 & 1 & \dots & 0 \\ \vdots & \vdots & \vdots & & \vdots \\ 0 & 0 & 0 & \dots & 1 \\ 0 & 0 & 0 & \dots & 0 \end{bmatrix} \text{ and } B = \text{col}(0,\dots,0,1)$$

then our problem reduces to determining a function $f \in W^{n,\infty}(a,b)$ for which $\{f^{(\nu)}(a)\}_{\nu=0}^{n-1} = X_0$, $\{f^{(\nu)}(b)\}_{\nu=0}^{n-1} = X_1$ and for which, $\|f^{(n)}\|_\infty$ is as small as possible. As we see from §6, this problem admits a unique solution f_0. The results of this chapter will show that this solution has the property that $f_0^{(n)}$ has constant modulus and at most n - 1 sign changes. More generally, we shall show that this same type of phenomenon continues to hold in a much more general setting. We shall use these and related results to show that the problem (6.1) admits a perfect spline solution f_0 with at most n knots between nodes.

To begin our discussion we note that X is a solution of (7.1) with initial condition $X(a) = X_0$ if and only if

$$(7.4) \qquad X(t) = \psi(t)\psi^{-1}(a)X_0 + \psi(t)\int_a^t \psi^{-1}(s)Bu(s)\,ds$$

where ψ is a fundamental matrix solution of the homogeneous system $X' = A(X)$. We assume throughout that the entries in the column $\psi(b)\psi^{-1}(s)B$ are linearly independent functions on every subset of [a,b] of positive measure. This will be the case, for example, if the coefficients of A are analytic on [a,b], in which case the fundamental matrix is analytic [7.3, p. 70] or when the first order linear system is equivalent to a single nth. order equation and B is the vector $[0,\dots,0,1]^t$. In this latter case, nonoscillation theorems for the solutions of the adjoint equation are relevant [7.3, p. 346].

For $\alpha \geq 0$, define

$$A_\alpha = \{X(b) \in \mathbb{R}^n : X(t) \text{ satisfies } (7.4) \text{ for some } u \in L^\infty,\ \|u\|_\infty \leq \alpha\}$$

The following proposition shows, in particular, that boundary points of A_α are attainable by bang-bang controls.

Proposition 7.1.

(1) A_α is a compact convex subset of $\mathbb{R}^n$;

(2) if X is a boundary point of A_α, then there is a nonzero vector $v \in \mathbb{R}^n$ with

$$X = \psi(b)\psi^{-1}(a)X_0 + \int_a^b \psi(b)\psi^{-1}(s)Bp(s)\,ds$$

where

$$p(s) = \alpha\ \operatorname{sgn}(v^t\psi(b)\psi^{-1}(s)B); \tag{7.5}$$

(3) If $\alpha > 0$, then the interior of A_α is non-empty; if X lies in the interior of A_α, then X lies in the interior of A_β for some $\beta < \alpha$.

Proof. (1) This is clear from the linearity of the map $u \rightarrow X(b)$ and the fact that this map is continuous with the weak-* topology on the α-ball of L^∞

(2) Since X is a point in the boundary of the compact convex set A_α, there is a support plane through X. Hence, there is a non-zero vector v with

$$v^t \cdot (X-Y) \geq 0$$

for all $Y \in A_\alpha$. Let $p \in L^\infty(a,b)$, $\|p\|_\infty \leq \alpha$, be a function associated with X by (7.4). Then for any $u \in L^\infty(a,b)$, $\|u\|_\infty \leq \alpha$, we have

$$\int_a^b [v^t\psi(b)\psi^{-1}(s)B]p(s)\,ds \geq \int_a^b [v^t\psi(b)\psi^{-1}(s)B]u(s)\,ds$$

Hence, p(s) must be given by (7.5) since, by hypothesis, $v^t\psi(b)\psi^{-1}(s)B$ cannot vanish on a set of positive measure.

(3) Let $u \in L^\infty(a,b)$, $\|u\| < \alpha$. Then X given by (7.4) for this u cannot be a boundary point of A_α by (2); hence, A_α has a non-empty interior.

Now let X lie in the interior of A_α for some $\alpha > 0$. Choose $\epsilon > 0$ so that $\{Y : \|X-Y\| < \epsilon\}$ is a subset of A_α. Let

$$M = \int_a^b \|\psi(b)\psi^{-1}(s)B\|\,ds.$$

We make the convention that the norm of a column of L^∞ functions is the Euclidean norm taken on the norms of the entries, with a like convention for members of $\mathbb{R}^n$. We claim that X lies in the interior of $A_{\alpha-\delta}$ for any $\delta < \min(\epsilon^2/4M,\alpha)$. If not, then there is a unit vector w with $w^t(X-Z) \geq 0$ for all $Z \in A_{\alpha-\delta}$. Let $X^* = X + (\epsilon/2)w$. Then

$$\|X^*-X\| = \epsilon/2 < \epsilon$$

so that $X^* \in A_\alpha$ and $\|X^*-Z\|^2 = \|(\epsilon/2)w+X-Z\|^2 = \epsilon^2/4 + \|X-Z\|^2 + \epsilon w^t(X-Z) \geq \epsilon^2/4$ for all $Z \in A_{\alpha-\delta}$. On the other hand, let $u \in L^\infty(a,b)$ be a function for which $\|u\| \leq \alpha$ and for which (7.4) yields $X(b) = X^*$. Define u_1 by

$$u_1(s) = \begin{cases} u(s) \text{ if } |u(s)| \leq \alpha - \delta \\ (\alpha-\delta)\,\mathrm{sgn}\, u(s) \text{ if } |u(s)| > \alpha - \delta \end{cases}$$

Let $X(b) = Z$ be the vector given by (7.4) for u_1 so that $Z \in A_{\alpha-\delta}$.

Then

$$\|Z-X^*\| \leq \int_a^b \|\psi(b)\psi^{-1}(s)B\|\|u(s)-u_1(s)\|_\infty ds$$
$$\leq M\delta < \epsilon^2/4$$

so that we have reached a contradiction. This completes the proof. □

Theorem 7.2. The problem (7.3) admits a solution u_0, where u_0 has the representation

$$u_0(s) = \alpha_0 \operatorname{sgn}(v^t\psi(b)\psi^{-1}(s)B). \tag{7.6}$$

Proof. The existence of a solution to (7.3) is straightforward and follows from Theorem 2.5; however, we shall sketch a direct proof. Let $\{u_n\}$ be a sequence of $L^\infty(a,b)$ functions for which (7.1) and (7.2) are solvable and for which $\alpha_0 = \lim\|u_n\|$. We may assume that $\{u_n\}$ converges weak-* to some function u_0 with $\|u_0\| \leq \alpha_0$. By (7.4) and weak-* convergence we clearly have

$$X_1 = \psi(b)\psi^{-1}(a)X_0 + \int_a^b \psi(b)\psi^{-1}(s)Bu_0(s)ds$$

and hence u_0 is a solution of (7.3). Thus $X_1 \in A_{\alpha_0}$ in the notation of Proposition 7.1. X_1 cannot be an interior point of A_{α_0} or else, by (3), X_1 would lie in $A_{\alpha_0-\delta}$ for some $\delta > 0$, contradicting the definition of α_0. Thus by (2) of Proposition 7.1, u_0 must be given by (7.6). □

Fix $\alpha > 0$ and let Y be the unique solution on [a,b] of $Y' = A(t)Y + B\alpha$ with $Y(a) = X_0$. Now define for $a \leq t \leq b$

$$\mathscr{A}(t) = \{X(b) : X \text{ satisfies (7.4) on } [t,b],\ X(t) = Y(t),\ \|u\|_\infty \leq \alpha\}$$

The following Proposition is proved much as 7.1 and we omit the proof.

Proposition 7.3.

(1) $\mathscr{A}(a) = A_\alpha$;

(2) For each $t \in [a,b]$, $\mathscr{A}(t)$ is a compact convex subset of $\mathbb{R}^n$;

(3) if X is a boundary point of $\mathscr{A}(t)$, then

$$X = \psi(b)\psi^{-1}(t)Y(t) + \int_a^b \psi(b)\psi^{-1}(s)Br(s)ds$$

where $\mathbf{r}(s) = \alpha \operatorname{sgn}(v^t\psi(b)\psi^{-1}(s)B)$ for $t \leq s \leq b$ and v is a non-zero vector in $\mathbb{R}^n$;

(4) if $t < b$, then $\mathscr{A}(t)$ has a non-empty interior;

(5) if $t < b$ and X lies in the interior of $\mathscr{A}(t)$, then X lies in the interior of $\mathscr{A}(t+\delta)$ for some $\delta > 0$.

Remark. It can be shown that the mapping $t \to \mathscr{A}(t)$ is continuous, where the range is topologized by the Hausdorff metric on compact subsets of $\mathbb{R}^n$.

Theorem 7.4. Let α_0 be as in (7.3). If $\alpha > \alpha_0$, then there is a function $q \in L^\infty(a,b)$ with $|q| = \alpha$ a.e. such that there is a solution X of (7.1) and (7.2) given by (7.4) with $u = q$. Further, there is a point $t_* \in [a,b]$ and a nonzero vector $v \in \mathbb{R}^n$ such that q is given by

$$q(t) = \alpha, \quad a \leq t < t_*$$
$$q(t) = \alpha \operatorname{sgn}(v^t\psi(b)\psi^{-1}(t)B), \quad t_* \leq t \leq b.$$

Proof. Fix $\alpha > \alpha_0$. By Theorem 7.2 we know that X_1 lies in A_{α_0} and X_1 also lies in the interior of A_α. Hence, Proposition 7.3 shows that X_1 lies in the interior of $\mathscr{A}(t)$ for $a \leq t \leq a + \delta$ for some $\delta > 0$. Let

$$t_1 = \text{lub}\{t : a \le t \le b,\ X_1 \in \text{interior } \mathscr{A}(t)\}$$

Clearly continuity implies $X_1 \in \mathscr{A}(t_1)$; (5) of Proposition 7.3 implies X_1 is a boundary point of $A(t_1)$. Hence, (3) of Proposition 7.3 gives a representation for X_1 for a certain function r on $[t_1,b]$ where $|r| = \alpha$ a.e. Define

$$q(s) = \begin{cases} \alpha & \text{on } [a,t_1] \\ r(s) & \text{on } [t_1,b] \end{cases}$$

and let X be a solution on [a,b] of $X' = AX + Bq$ with $X(a) = X_0$. Then

$$\begin{aligned} X(b) &= \psi(b)\psi^{-1}(a)X_0 + \int_a^b \psi(b)\psi^{-1}(s)Bq(s)\,ds \\ &= \psi(b)\psi^{-1}(a)X_0 + \alpha\int_a^{t_1} \psi(b)\psi^{-1}(s)B\,ds \\ &\quad + X_1 - \psi(b)\psi^{-1}(t_1)Y(t_1) \\ &= \psi(b)\psi^{-1}(a)X_0 + \alpha\int_a^{t_1} \psi(b)\psi^{-1}(s)B\,ds \\ &\quad + X_1 - \psi(b)\psi^{-1}(t_1)[\psi(t_1)\psi^{-1}(a)X_0 \\ &\qquad + \alpha\int_a^{t_1} \psi^{-1}(t_1)\psi^{-1}(s)B\,ds] \\ &= X_1 \end{aligned}$$

This proves Theorem 7.4. □

7.2. Linear Systems With Multipoint Boundary Conditions

We shall utilize Theorem 7.4 in this section to obtain optimal bang-bang controls for solutions of linear systems satisfying multipoint boundary conditions. Suppose then that we are given points

$$a \leq t_1 < t_2 < \ldots \leq t_r \leq b$$

and mixed multipoint boundary conditions determined by a full rank $m \times rn$ matrix D, $m \leq rn$. We may formulate the generalization of (7.2) as follows.

$$D[X(t_1),\ldots,X(t_r)]^t \in K \tag{7.7}$$

Here K is a compact convex subset of $\mathbb{R}^m$ and X has n absolutely continuous components with essentially bounded derivatives.

Theorem 7.5. Consider the extremal problem

$$\alpha = \inf\{\|u\|_\infty : (7.1) \text{ is solvable subject to } (7.7)\}. \tag{7.8}$$

Then (7.8) admits a solution u_0, $|u_0| = \alpha$ a.e. on $[a,b]$ whose restriction to each subinterval $[a,t_1]$, $[t_1,t_2],\ldots,[t_r,b]$ has the structure of the function q in Theorem 7.4.

Proof. The existence of an optimal control u, satisfying (7.1) and (7.7), and solving (7.8), is a consequence of Theorem 2.5. Now we consider a (vector) function X satisfying (7.1) and (7.7) for this optimal control. Let Z be the (row) vector $Z = [X(a),X(t_1),\ldots,X(t_r),X(b)]$ if $a < t_1 < t_r < b$, with the obvious deletions if $a = t_1$ or $t_r = b$. On any given subinterval determined by $a,t_1,\ldots,t_r,b$, the corresponding two point problem (7.1), (7.2) has an infimal constant $\alpha_0 \leq \alpha$, where α is given by (7.8). Here we are using for boundary values in (7.2) the appropriate entries of Z. Theorem 7.4 now gives the conclusion. Note that the absolutely continuous (vector) function $\tilde{X}$ so determined must satisfy (7.7) since X does. $\square$

7.3 Applications to Perfect Spline Functions

<u>Theorem 7.6</u>. Let $L = D^n + \sum_{j=1}^{n} a_j D^j$ be a non-singular linear differential operator on [0,1] with $a_j \in C^j[0,1]$, let $0 \le x_1 < \ldots < x_r \le 1$ be prescribed points, let the L_{ij} be the functionals given in section 6.3 and let α be defined by (6.1). Then without any consistency assumptions (6.1) admits a solution f_0 for which $|f_0^{(n)}| = \alpha$ a.e. on (0,1), and Lf has only a finite number of sign changes on [0,1]. If the null space of L* is spanned by a Chebyshev system, then Lf has n or fewer sign changes in (x_i, x_{i+1}) for $i = 1,\ldots,r-1$, with a similar statement for $(0,x_1)$, $(x_r,1)$.

Proof. Take

$$A = \begin{bmatrix} 0 & 1 & 0 & \ldots & 0 \\ 0 & 0 & 1 & \ldots & 0 \\ \vdots & \vdots & \vdots & & \vdots \\ 0 & 0 & 0 & \ldots & 1 \\ -a_0 & -a_1 & & & -a_{n-1} \end{bmatrix},$$

$B = (0,\ldots,0,1)^t$ and the matrix D to be a block diagonal matrix whose (nonsquare) blocks have rows consisting of the coefficients of the L_{ij}; here the blocks are indexed by i and the block rows by j. Also, $m = \sum_{i=1}^{r} k_i$ represents the number of rows of D. Now by Theorem 7.5 there is a solution of (7.8). By the choice of A and B, u = Lf for a solution f of (6.1). Moreover, u may be chosen so that $|u| = \alpha$ a.e. and so that, on each subinterval, determined by

$0, x_1, \ldots, x_r, 1$, $u = \alpha$ followed by $u(s) = \alpha \operatorname{sgn}(v^t \psi(\lambda) \psi^{-1}(s) B)$, where λ is the right endpoint of the given subinterval. Since $\psi(\lambda)\psi^{-1}(s)B$ represents a column of null solutions of L^*, whose linear span contains functions vanishing only on a finite point set in $[0,1]$ if nonzero, it follows that Lf has only a finite number of discontinuties on $[0,1]$. If the null space of L^* is spanned by a Chebyshev system then $\operatorname{sgn}(v^t \psi(\lambda) \psi^{-1}(s) B)$ has at most $n - 1$ discontinuities on its subinterval, so that u has at most n discontinuities. This concludes the proof. □

Remarks. The ideas of this chapter are taken from the paper of McClure [7.6]; Proposition 7.1 is essentially from the book of Hermes and LaSalle [7.4]. We emphasize that the bang-bang phenomena of Theorem 7.5, which is the central result of the chapter, hold for very general linear systems and multipoint boundary inequalities. The authors believe that McClure's paper is a significant one, in that it clearly exploits the well-understood concepts of control theory in the setting of L^∞ minimization problems. Schoenberg [7.8] recognized the connection between the time optimal control problem and a special two point Hermite interpolation problem of minimal norm, and gave an explicit solution following Louboutin [7.5]. The bang-bang principle in control theory is well understood and was set forth in the book of L. Pontryagin and colleagues [7.7] as a consequence of the maximum principle. An especially revealing demonstration, bypassing the maximum principle and utilizing extreme points, is presented in Berkovitz [1.2]. A derivation is also presented in Lions [1.8].

Theorem 7.6 is the pivotal application in [7.6]. The existence of piecewise perfect spline functions as the solution of a sequence of explicitly describable intrinsic L^∞ minimization problems was demonstrated in [6.1]. Theorem 7.6, for the case $L = D^n$, was essentially obtained in [7.1]. This proof utilized the core interval of uniqueness, on which all solutions are perfect spline functions under complete consistency of the functionals, and employed local extension arguments. In the present setting, however, we note that Theorem 7.6 is valid in substance with no assumptions on L or the functionals and holds for convex constraints as well. This accrues from the logical order of proof, viz., existence is demonstrated first, followed by bang-bang interpolation. The fundamental role of [7.1] lies in the construction of the bang-bang interpolant via the solution of nonlinear systems of algebraic equations; existence questions were resolved in the authors' earlier papers [1.5, 6.1].

For completeness, we should also cite the paper of G. Glaeser [7.2] and the dissertation of P. Smith [7.9]. These played important parts in the L^∞ interpolation-minimal norm investigations. We shall return to perfect splines in the sequel.

We make one final remark. The conclusion of Theorem 7.6 can be strengthened if the functionals are completely consistent and the null space of L* is spanned by a Chebyshev system. In this case, Theorem 7.2 implies that on each subinterval (x_i, x_{i+1}) of the core interval of uniqueness, the solution has at most n - 1 knots.

REFERENCES

7.1 S. D. Fisher and J. W. Jerome, "Perfect spline solutions to L^∞ extremal problems," J. Approximation Theory, 12 (1974), 78-90.

7.2 G. Glaeser, "Prolongement extrèmal de fonctions differentiables," Publ. Sect. Math. Faculte des Sciences Rennes, Rennes, France, 1967.

7.3 P. Hartman, Ordinary Differential Equations, Wiley, New York, 1964.

7.4 H. Hermes and J. P. LaSalle, Functional Analysis and Time Optimal Control, Academic Press, New York, 1969.

7.5 R. Louboutin, "Sur une bonne partition de l'unite," in, Le Prolongateur de Whitney, Vol. II (G. Glaeser, ed.), Universite de Rennes, Rennes, France, 1967.

7.6 D. E. McClure, "Perfect spline solutions of L_∞ extremal problems by control methods," J. Approximation Theory, to appear.

7.7 L. S. Pontryagin, V. G. Boltyanskii, R. V. Gamkrelidze and E. F. Mischenko, The Mathematical Theory of Optimal Processes, Interscience, New York, 1962.

7.8 I. J. Schoenberg, "The perfect B-splines and a time-optimal control problem," Israel J. Math., 10 (1971), 261-274.

7.9 P. W. Smith, "$W^{r,p}(\mathbb{R})$-splines," Dissertation, Purdue University, Lafayette, Indiana, 1972.

V. LAGRANGE MULTIPLIER SOLUTIONS WITH INEQUALITIES

§8. A General Theorem of Kuhn-Tucker Type

In this part we consider some extensions of the variational problems of §3 where the constraints are allowed to be non-linear. As we show these more general problems can be handled in much the same way as linear constraints. In §9 we discuss in detail a particular problem of this type and elaborate the nature of its solution. This section contains a general theorem and an illustrative example. The general theorem is a concretization of a result of Luenberger [8.1, Theorem 1, p. 249] concerning Lagrange multipliers. Our method of proof differs from that of [8.1] and for completeness we present it.

Theorem 8.1. Let T be a Frechet differentiable operator from a Banach space X into $L^p(\Omega,\mu)$, $1 \le p < \infty$ where Ω is a σ-finite measure space; let $\ell_1,\dots,\ell_m$ be Frechet differentiable functionals on X. Let $\underline{r} = (r_1,\dots,r_m) \in \mathbb{R}^m$ and set

$$(8.1)\qquad \begin{aligned} U_{\underline{r}} &= \{x \in X : \ell_i(x) \le r_i,\ 1 \le i \le m\} \\ \alpha_{\underline{r}} &= \inf\{\|Tx\|_p : x \in U_{\underline{r}}\} \end{aligned}$$

Let $x_0 \in U_{\underline{r}}$ be a solution of (8.1), let L be the Frechet derivative of T at x_0, and let L^a be the adjoint of L, $L^a : L^q \to X^*$, $1/q + 1/p = 1$. Then there are scalars $\beta_1,\dots,\beta_m$, all of which are less than or equal to 0, such that

$$(8.2)\qquad L^a(u|Tx_0|^{p-1}) = \sum_1^m \beta_j \ell_j'$$

where $u = \operatorname{sgn} Tx_0$ and ℓ_j' is the Frechet derivative of ℓ_j, $1 \le j \le m$, provided the cone

$$V = \{x \in X : \ell_j'(x) < 0,\ j = 1,\ldots,m\}$$

is non-void.

Proof. If $x \in V$, then $x_0 + \epsilon x \in \underline{U_r}$ for sufficiently small $\epsilon > 0$ so that $\|Tx_0+\epsilon x)\|_p \geq \alpha_r$. Expanding we find that

$$(8.3) \qquad 0 \leq \int_\Omega u|Tx_0|^{p-1}Lx\, d\mu$$

Hence, whenever $x \in V$, then $\int_\Omega L^a(u|Tx_0|^{p-1})x\, d\mu \geq 0$. Let $C = \{\sum_1^m \beta_j\ell_j' : \beta_j \leq 0,\ 1 \leq j \leq m\}$; then C is a weak-* closed convex cone in X* and hence has the property that it is the conjugate cone of the cone,

$$M = \{x \in X : x^*(x) \geq 0 \text{ for all } x^* \in C\},$$

i.e., $C = \{x^* \in X^* : x^*(x) \geq 0 \text{ for all } x \in M\}$. Now $V \subset M$ and M is the weak closure of V. It follows that C is the conjugate cone of V and, by (8.3), $u|Tx_0|^{p-1} \in C$. □

Remarks. The result that C is the conjugate cone of M is a consequence of the proof of Theorem 4.62-A of [2.8] where a reflexivity result for annihilators of subspaces is obtained.

It is possible to expand the conclusion of Theorem 8.1 to include the assertion that $\beta_j = 0$ if $\ell_j(x_0) = r_j$, $j = 1,\ldots,m$, (cf. [8.1, p. 249]). This gives, then, a result of Kuhn-Tucker type for the stationary principle enunciated in (8.2).

Example 8.1. Let points $\{x_i\}_{i=1}^m$ be given in [0,1], $0 \leq x_1 < x_2 < \ldots < x_m \leq 1$, and let real numbers $\{y_i\}_{i=1}^m$ be given. Take $X = W^{2,2}[0,1]$, let I_1 and I_2 be subsets of $\{1,\ldots,m\}$, let M be a positive number and let U consist of all functions in X satisfying

$$(8.4)\qquad \begin{aligned}&\text{(i)}\quad f(x_i) \le y_i,\ i \in I_1\\ &\text{(ii)}\quad f(x_i) \ge y_i,\ i \in I_2\\ &\text{(iii)}\quad \int_0^1 |f'|^6 \le M\end{aligned}$$

(We assume M is chosen so that U is non-empty.) Let

$$Tf = f''(1+(f')^2)^{-5/4}$$

and consider the minimization problem

$$(8.5)\qquad \alpha = \inf_{f \in U} \|Tf\|_2$$

Geometrically, the problem is to find the graph passing above certain points and below certain others whose curvature is as small as possible. The condition (8.4iii) is included because it will insure that (8.5) has a solution as we show below.

If $f \in U$ and $\|Tf\|_2 \le 2\alpha$, then with $p = 12/11$,

$$\int_0^1 |f''|^p = \int_0^1 |f''|^p(1+(f')^2)^{5p/4}(1+(f')^2)^{-5p/4}$$

$$\le \|Tf\|_2^{2/p}\left(\int_0^1 (1+(f')^2)^3\right)^{1/2}$$

$$\le C$$

where C depends only on α and M. Hence, if $\{f_n\}$ is a sequence of elements of U with $\|Tf_n\|_2 \to \alpha$, then by extracting appropriate subsequences we may assume that Tf_n converges weakly in $L^p(0,1)$ to $f''/(1+(f')^2)^{5/4}$ where $f'' \in L^p(0,1)$. But we can also assume that Tf_n converges weakly in L^2 to a function g satisfying $\|g\|_2 \le \alpha$. Since

$f' \in L^\infty(0,1)$ we find that $g = f''(1+(f')^2)^{-5/4}$ and $f'' \in L^2(0,1)$. Thus, f is a solution to (8.5).

Now provided the cone V of Theorem 8.1 is nonempty we may apply Theorem 8.1 with $\ell_1,\ldots,\ell_m$ the (linear) functionals of evaluation at x_i (or minus evaluation at x_i in case $i \in I_2$) and $\ell_{m+1}(f) = \int_0^1 |f'|^6$. The Frechet derivative of T at f_0 is

$$Lg = F_1 g' + F_2 g''$$

where

$$F_1 = -\frac{5}{2} f_0''(1+(f_0')^2)^{-9/4} f_0'$$

and

$$F_2 = (1+(f_0')^2)^{-5/4}$$

The Frechet derivative of ℓ_i is ℓ_i for $i = 1,\ldots,m$ and the Frechet derivative of ℓ_{m+1} is

$$\ell_{m+1}'(g) = 6 \int_0^1 (f_0')^5 g'$$

Hence, if $v \in C^\infty$ and $v(x_i) = 0$ for $i = 1,\ldots,m$ and if the cone $V = \{f \in W^{2,2}(0,1) : \ell_j'(f) < 0,\ j = 1,\ldots,m+1\}$ is nonempty, then by Theorem 8.1,

$$\int_0^1 (Lv)Tf_0 = \lambda \ell_{m+1}'(v).$$

It follows that in each subinterval (x_i, x_{i+1}), $i = 1,\ldots,m-1$, f_0 satisfies the differential equation

$$2D^2 \frac{f_0''}{(1+f_0'^2)^{5/2}} + 5D \frac{f_0''^2 f'}{(1+f_0'^2)^{7/2}} - \lambda f_0'^4 = 0$$

where $\lambda \leq 0$. Hence, f_0 is C^∞ in each (x_i, x_{i+1}).

Remarks. Example 8.1 is of course the problem of minimum curvature for graphs. The condition (8.4iii) is a localizing condition and is a substitute for the hypothesis of a bounded minimizing sequence. The reader interested in pursuing a general treatment of Lagrange multipliers is referred to [8.1] where numerous examples are presented. A version of Example 8.1 was first considered in [1.5].

REFERENCES

8.1 D. G. Luenberger, Optimization by Vector Space Methods, Wiley, New York, 1969.

§9. Stable and Unstable Elastica Equilibrium and the Problem of Minimum Curvature

In this section we explore in some detail the properties of the solutions of the problems first presented in Examples 1.1 and 1.2.

9.1 Characterizations of Solutions in the case $p = 2$

In this section we assume that L is a fixed length parameter and that a finite set $\mathscr{P}$ of points (x_i, y_i), $i = 1,\ldots,m$ is prescribed which admits a smooth interpolating curve, with continuous turning tangent and square integrable curvature, of length not exceeding L.

Let $W^{2,2}(0,L)$ denote the usual Sobolev class of real functions x such that $x, \dot{x}$ are absolutely continuous on $[0,L]$ and $\ddot{x} \in L^2(0,L)$. Let W be the product space

$$W = W^{2,2}(0,L) \times W^{2,2}(0,L)$$

and define the functional ℓ on W by

$$\ell(x,y) = \int_0^L \sqrt{\dot{x}^2 + \dot{y}^2}.$$

Also, we define the functional θ on W to be the square of the L^2 norm of the curvature of the curve $t \to (x(t),y(t))$, $0 \le t \le L$. This can be given explicitly by

$$\theta(x,y) = \int_0^L (\dot{x}\ddot{y}-\dot{y}\ddot{x})^2(\dot{x}^2+\dot{y}^2)^{-5/2}.$$

We say a point (x,y) in W is admissible if $\mathscr{P}$ is a subset of $\{(x(t),y(t)) : 0 \le t \le L\}$. The basic problem considered, then, in this section is to characterize solutions of the extremal problem

$$(9.1)\quad \alpha(L) = \inf\{\theta(x,y) : \ell(x,y) \leq L,\ (x,y) \text{ admissible}\}.$$

As shown in Example 1.1, solutions of this problem exist. Note that $\alpha(L)$ is a monotone decreasing function of L.

Theorem 9.1. Let (x,y) be a solution of (9.1) parametrized by the arc-length variable t. Let $0 = t_1 < t_2 < \ldots < t_m \leq L$ be points with $x(t_i) = x_i$, $y(t_i) = y_i$ for $i = 1,\ldots,m$. (This can be accomplished by renumbering the points of $\mathscr{P}$, if necessary.) Then each of x and y is C^∞ in (t_i,t_{i+1}), $i = 1,\ldots,m - 1$, and C^2 on $[0,L]$ with $\ddot{x}(0) = \ddot{x}(t_m) = \ddot{y}(0) = \ddot{y}(t_m) = 0$. Further, there is a nonnegative constant λ and there are constants c_i, d_i such that

$$(9.2)\qquad \begin{aligned} 3\varkappa^2\dot{x} + 2\dddot{x} &= \lambda\dot{x} + c_i \\ 3\varkappa^2\dot{y} + 2\dddot{y} &= \lambda\dot{y} + d_i \end{aligned}$$

on (t_i,t_{i+1}) for $i = 1,\ldots,m - 1$ where $\varkappa$ is the curvature and differentiation is taken with respect to arc length. The system (9.2) implies that the equations

$$(9.3)\qquad \ddot{\varkappa} + \frac{\varkappa^3}{2} - \frac{\lambda}{2}\varkappa = C_i$$

hold on each interval (t_i,t_{i+1}) for certain constants C_i, $i = 1,\ldots,m - 1$. Furthermore, if $t_m < L$ then (9.2) and (9.3) hold with $\lambda = 0$.

Two technical lemmas, which make use of variational inequalities, are required for the proof of Theorem 9.1.

Lemma 9.2. Suppose f, $g \in L^2(0,1)$ and

$$\int_0^1 f\ddot{a} \geq 0 \text{ implies } \int_0^1 g\ddot{a} \geq 0$$

for any $a \in C_0^\infty(0,1)$. Then $g(t) = \lambda f(t) + c_0 + c_1 t$ a.e. in $[0,1]$ where $\lambda \geq 0$ and c_0, c_1 are real numbers.

Proof. The lemma clearly holds if f is an affine function of t so we may suppose f is not affine. Let
$S = \{\lambda f + c_0 + c_1 t : \lambda \geq 0,\ c_0,\ c_1 \in \mathbb{R}^1\}$; then S is a closed convex cone in L^2 and we wish to show that $g \in S$. If not, by the separation theorem there is an L^2 function p with $\int_0^1 pg = \delta < 0$ and $0 \leq \int_0^1 sp$ for all $s \in S$. Hence, $\int_0^1 pf \geq 0$ and $\int_0^1 p(t)\,dt$ $= \int_0^1 tp(t)\,dt = 0$. If $\int_0^1 pf = 0$, then since the C_0^∞ functions β which are orthogonal to 1, t, f are L^2 dense in the set of L^2 functions which are orthogonal to 1, t, f, there is a C_0^∞ function β approximating p with $\int_0^1 \beta g \leq \delta/2$ and

$0 = \int_0^1 \beta = \int_0^1 t\beta(t)\,dt = \int_0^1 f\beta$; [9.3; Theorem 0]. If $\int_0^1 Pf > 0$, then by the same argument there is a C_0^∞ function β with $0 = \int_0^1 \beta = \int_0^1 t\beta(t)\,dt$ and $\int_0^1 \beta g \leq \delta/2$, $\int_0^1 \beta f > 0$. The conditions $0 = \int_0^1 \beta = \int_0^1 t\beta(t)\,dt$ imply that $\beta = \ddot{\alpha}$ for some $\alpha \in C_0^\infty$; hence, $\int_0^1 \ddot{\alpha} f \geq 0$ so that the hypothesis implies $0 \leq \int \ddot{\alpha} g = \int \beta g \leq \delta/2 < 0$, a contradiction. □

Corollary 9.3. Suppose $f, g \in L^2(0,1)$ and

$$\int_0^1 f\ddot{a} > 0 \text{ implies } \int_0^1 g\ddot{a} \geq 0$$

for any $a \in C_0^\infty(0,1)$. Then $g(t) = \lambda f(t) + c_0 + c_1 t$ a.e. in $(0,1)$ where $\lambda \geq 0$ and c_0, c_1 are real numbers.

Proof. If $\int_0^1 f\ddot{a} = 0$, replace $\ddot{a}$ by $\ddot{a} + \epsilon\ddot{b}$ where $\int_0^1 \ddot{b}f > 0$ and let ϵ decrease to 0. □

Lemma 9.4. Let $f \in L^2(0,1)$, let $r \in \mathbb{R}$, $0 < c < 1$, and set $U = \{u \in C_0^\infty(0,1) : \dot{u}(c) = r, \text{ and } u(c) = \int_0^1 \dot{u}f = 0\}$. Then there is a sequence $\{u_n\}$ of elements of U with $\|\dot{u}_n\|_2 \to 0$.

Proof. Let $V = \{v \in C_0^\infty(0,1) : \int_0^1 v = \int_0^c v = \int_0^1 vf = 0$ and $v(c) = r\}$. Then the conclusion of the lemma is that 0 is in the L^2 closure of the convex set V. First we show that V is non-empty. Let $w \in C_0^\infty$, $w(c) = r$. Since the set of C_0^∞ functions which vanish at c is dense in L^2, there is such a C_0^∞ function v_1 with $\int_0^1 v_1 = -\int_0^1 w$, $\int_0^c v_1 = -\int_0^c w$, and $\int_0^1 v_1 f = -\int_0^1 wf$ [9.3; Theorem 0]. Then $v = v_1 + w \in V$. If 0 is not in the L^2 closure of V, there is an L^2 function g with $0 < \delta \leq \int_0^1 vg$ for all $v \in V$. Fix $v_0 \in V$ and choose any $w \in C_0^\infty$ with $\int_0^1 w = \int_0^c w = \int_0^1 wf = w(c) = 0$. If $\lambda \in \mathbb{R}^1$, then $v_0 + \lambda w \in V$ and so $\delta \leq \int_0^1 (v_0+\lambda w)g$, and hence $\int_0^1 wg = 0$. Another application of [9.3; Theorem 0] shows that the L^2 closure of the set of C_0^∞ functions w satisfying the four conditions: $0 = \int_0^1 w = \int_0^c w = \int_0^1 wf = w(c)$ is precisely the set of

L^2 functions w satisfying the first three of these conditions. Hence, $g = A_1 + A_2\chi + A_3 f$ where χ is the characteristic function of $[0,c]$. Hence, for $v \in V$, $0 < \delta \leq \int_0^1 gv = 0$ which is a contradiction. □

<u>Proof of Theorem 9.1</u>. Let t be the arc-length parameter for the solution (x,y). Note that $\dot{x}^2 + \dot{y}^2 \equiv 1$ and also that

$\theta(x,y) = \int_0^L (\ddot{x}^2+\ddot{y}^2)$ in this parametrization. (We are also assuming that the parameter interval is $[0,L]$; this involves no loss of generality since we may always extend each of x and y linearly to all of $[0,L]$.) Let $(a,b) \in W$ with $a(t_i) = b(t_i) = 0$ for $i = 1, \ldots, m$. Then $(x+\epsilon a, y+\epsilon b)$ satisfies the interpolation conditions and, for ϵ sufficiently small,

$$\ell(x+\epsilon a, y+\epsilon b) = \ell(x,y) + \epsilon\ell'(a,b) + O(\epsilon^2) \tag{9.4}$$

where $\ell'(a,b) = \int_0^L (\dot{a}\dot{x}+\dot{b}\dot{y})\,dt$ is the Frechet derivative of ℓ at (x,y). For θ we have

$$\begin{aligned}\theta(x+\epsilon a, y+\epsilon b) &= \int_0^L [(\dot{x}+\epsilon\dot{a})(\ddot{y}+\epsilon\ddot{b})-(\ddot{x}+\epsilon\ddot{a})(\dot{y}+\epsilon\dot{b})]^2 \\ &\qquad \cdot [(\dot{x}+\epsilon\dot{a})^2+(\dot{y}+\epsilon\dot{b})^2]^{-5/2}\,dt \\ &= \theta(x,y) + \epsilon\theta'(a,b) + O(\epsilon^2)\end{aligned}$$

where

$$\theta'(a,b) = \int_0^L [\dot{x}\ddot{y}-\ddot{x}\dot{y}][2(\dot{a}\ddot{y}-\ddot{a}\dot{y}+\dot{x}\ddot{b}-\ddot{x}\dot{b})-5(\dot{a}\dot{x}+\dot{b}\dot{y})(\dot{x}\ddot{y}-\ddot{x}\dot{y})]\,dt.$$

Making use of the facts that $\dot{x}^2 + \dot{y}^2 = 1$ and hence $\dot{x}\ddot{x} + \dot{y}\ddot{y} = 0$ we easily obtain the formula

$$(9.5) \qquad \theta'(a,b) = \int_0^L [2(\ddot{a}\ddot{x}+\ddot{b}\ddot{y})-3\kappa^2(\dot{a}\dot{x}+\dot{b}\dot{y})]dt$$

where $\kappa^2 = \ddot{x}^2 + \ddot{y}^2$ is the square of the curvature.

Fix i, $1 \le i \le m-1$. Suppose a, b $\in C_0^\infty(t_i,t_{i+1})$ and $\ell'(a,b) < 0$. Suppose $\theta'(a,b) = \eta < 0$; then the curve $(x+\epsilon a, y+\epsilon b)$ has length $L - \delta(\epsilon) < L$ for ϵ sufficiently small, but

$$\begin{aligned}\alpha(L) \le \alpha(L-\delta(\epsilon)) &\le \theta(x+\epsilon a, y+\epsilon b)\\ &= \theta(x,y) + \epsilon\theta'(a,b) + O(\epsilon^2)\\ &= \alpha(L) + \epsilon\eta + O(\epsilon^2)\\ &< \alpha(L)\end{aligned}$$

when ϵ is sufficiently small. This contradiction shows that $\theta'(a,b) \ge 0$. Now take $b \equiv 0$. We learn that, for any

$a \in C_0^\infty(t_i,t_{i+1})$, $\int_{t_i}^{t_{i+1}} \dot{a}\dot{x} < 0$ implies that $\int_{t_i}^{t_{i+1}} (2\ddot{a}\ddot{x}-3\kappa^2\dot{a}\dot{x}) \ge 0$.

Equivalently, $\int_{t_i}^{t_{i+1}} \ddot{a}x > 0$ implies that $\int_{t_i}^{t_{i+1}} 2\ddot{a}\ddot{x} + \ddot{a}F \ge 0$

where $\dot{F} = 3\kappa^2\dot{x}$. Now, by Corollary 9.3, we must have

$$2\ddot{x} + F = \lambda_i x + \tilde{c}_i + c_i t, \quad t \in (t_i,t_{i+1})$$

where $\lambda_i \ge 0$ and $\tilde{c}_i, c_i$ are scalars. This immediately implies that $\ddot{x}$ is absolutely continuous with $\dddot{x}$ in $L^2(t_i,t_{i+1})$ and satisfies on (t_i,t_{i+1})

$$(9.6) \qquad 2\dddot{x} + 3\kappa^2\dot{x} = \lambda_i\dot{x} + c_i, \; i = 1,\dots,m-1.$$

Taking $a \equiv 0$ and employing the identical argument we find that y satisfies

$$(9.7) \qquad 2\dddot{y} + 3\kappa^2\dot{y} = \mu_i\dot{y} + d_i, \; i = 1,\dots,m-1.$$

A recursion argument applied to (9.6) and (9.7) implies that x and y are in $C^\infty(t_i, t_{i+1})$.

To prove the continuity of $\ddot{x}$ at the points t_i we make use of Lemma 9.4 and the fact that $\ell'(a,b) < 0$ implies $\theta'(a,b) \geq 0$. First, if $2 \leq i \leq m-1$, then taking $b \equiv 0$ and $a \in C_0^\infty(t_{i-1}, t_{i+1})$ with $a(t_i) = 0$, $\dot{a}(t_i) = r$, we find that whenever

$\int_{t_{i-1}}^{t_{i+1}} \dot{a}\dot{x} < 0$, then

$$0 \leq \int_{t_{i-1}}^{t_{i+1}} (2\ddot{a}\ddot{x} - 3\kappa^2\dot{a}\dot{x}) = -\int_{t_{i-1}}^{t_{i+1}} (2\dot{a}\dddot{x} + 3\kappa^2\dot{a}\dot{x}) + 2r[\ddot{x}(t_i^-) - \ddot{x}(t_i^+)]$$

By a continuity argument similar to that employed in the proof of the Corollary 9.3, this same inequality holds if $\int_{t_{i-1}}^{t_{i+1}} \dot{a}\dot{x} \leq 0$. By Lemma 9.4, the first term on the right hand side can be made arbitrarily small for any choice of r and, in particular, $r = \pm 1$. Thus $\ddot{x}(t_i^-) = \ddot{x}(t_i^+)$. A similar argument with a simpler version of Lemma 2 shows that $\ddot{x}(0) = \ddot{x}(t_m) = 0$. Similar arguments hold for $\ddot{y}$.

We now show that the constants λ_i and μ_i in (9.6) and (9.7) are equal to a single constant λ. We first consider the case that $\dot{x}$ is constant on $[0,L]$. If so, then (9.6) implies that the λ_i's are arbitrary if $\dot{x} \equiv 0$ and, otherwise, κ^2 is constant on each interval (t_i, t_{i+1}).

Since κ^2 is continuous on $[0,L]$, and has value 0 at 0 and L, we find that $\kappa^2 \equiv 0$. Thus the solution (x,y) is a straight line and $\lambda = 0$ in this case. We now assume that $\dot{x}$ is not constant on $[0,L]$.

Let $J_i = (t_i, t_{i+1})$ for $i = 1,\ldots,m-1$. Let $n_1,\ldots,n_k$ be those indices such that $\dot{x}$ is not constant on J_{n_i}, $i = 1,\ldots,k$. For simplicity let $L_i = J_{n_i}$, $i = 1,\ldots,k$. We know from an earlier portion of the proof of Theorem 9.1 that, if $(a,b) \in W$ and $a(t_i) = b(t_i) = 0$ for $i = 1,\ldots,m$, then $\ell'(a,b) < 0$ implies $\theta'(a,b) \geq 0$. Making use of equations (9.6) and (9.7), valid on J_i, we find that

$$(9.8) \quad \sum_1^k \Big[\int_{L_i} \dot{a}\dot{x} + \dot{b}\dot{y}\Big] < 0 \text{ implies } \sum_1^k \Big[\lambda_{n_i} \int_{L_i} \dot{a}\dot{x} + \mu_{n_i} \int_{L_i} \dot{b}\dot{y}\Big] \geq 0.$$

Let U be the linear space of C^∞ functions u on $[0,L]$ with $u(t_i) = 0$ for $i = 1,\ldots,m$ and consider the mapping T of U into $\mathbb{R}^k$ given by

$$Tu = \Big\{\int_{L_i} \dot{u}\dot{x}\Big\}_{i=1}^{k}.$$

TU is a subspace of $\mathbb{R}^k$. If it is not all of $\mathbb{R}^k$, then there are scalars $\beta_1,\ldots,\beta_k$ not all zero with

$$0 = \sum_{i=1}^{k} \beta_i \int_{L_i} \dot{u}\dot{x}, \text{ for all } u \in U.$$

By choosing $u \in C^\infty$ with compact support in L_{i_0} we find that $\beta_{i_0} = 0$ since $\dot{x}$ is not constant on L_i; this holds for $1 \leq i_0 \leq k$ so that we reach a contradiction. Hence, TU is all of $\mathbb{R}^k$.

Now use (9.8) with $b \equiv 0$; we find that

$$\sum_1^k v_i < 0 \text{ implies } \sum_1^k \lambda_{n_i} v_i \geq 0$$

for any $(v_1,\ldots,v_k) \in \mathbb{R}^k$. It follows immediately that $\lambda_{n_1} = \ldots = \lambda_{n_k} = \lambda$. On the other hand, on an interval J_i on which $\dot{x}$ is constant, the equation $2\dddot{x} + 3\kappa^2\dot{x} = \lambda_i\dot{x} + c_i$ shows that either λ_i is arbitrary if $\dot{x} \equiv 0$ or that κ^2 is constant and an adjustment of c_i allows us to assume $\lambda_i = \lambda$. Hence, $\lambda_1 = \ldots = \lambda_{m-1} = \lambda$. An identical argument shows that $\mu_1 = \ldots = \mu_{m-1} = \mu$. It only remains to show that $\mu = \lambda$. We again use (9.8), which now takes the form,

$$\sum_1^k v_i + \sum_1^\ell w_i \leq 0 \text{ implies } \lambda \sum_1^k v_i + \mu \sum_1^\ell w_i \geq 0$$

for any $(v_1,\ldots,v_k) \in \mathbb{R}^k$ and $(w_1,\ldots,w_\ell) \in \mathbb{R}^\ell$. This clearly implies $\lambda = \mu$.

Now we derive the differential equation (9.3) which the curvature κ satisfies. Our starting point is the system of equations (9.2). Differentiate the first equation in (9.3) and then multiply by $\dot{y}$. Next, differentiate the second equation in (9.2) and then multiply by $-\dot{x}$. Add the resulting equations to obtain

$$(9.9) \qquad -3\kappa^3 + 2(x^{(4)}\dot{y} - y^{(4)}\dot{x}) = -\lambda\kappa.$$

Here we have used the fact that $\kappa^2 = \ddot{x}^2 + \ddot{y}^2$. A simple computation shows that

$$(9.10) \qquad x^{(4)}\dot{y} - y^{(4)}\dot{x} = \ddot{x}\dddot{y} - \dddot{x}\ddot{y} - \ddot{\kappa}.$$

However, for some constant γ_i,

$$(9.11) \qquad \ddot{x}\dddot{y} - \dddot{x}\ddot{y} = \varkappa^3 + \gamma_i .$$

To see this, differentiate the first equation of (9.2) and then multiply by $\ddot{y}$. Next, differentiate the second equation of (9.2) and then multiply by $\ddot{x}$. Add the resulting equations to obtain the differentiated form of (9.11). (9.3) is then a consequence of (9.9), (9.10) and (9.11).

Finally, suppose that $t_m < L$. Then clearly $\dot{x}$ and $\dot{y}$ are constant on $[t_m, L]$. (Otherwise, we could decrease $\alpha(L)$ by considering only $[0, t_m]$.) On $[t_m, L]$ we then have the equations

$$2\ddddot{x} + 3\varkappa^2\dot{x} = 0$$

and

$$2\ddddot{y} + 3\varkappa^2\dot{y} = 0.$$

Now if $(a,b) \in W(0,t_m)$ with $a(t_i) = b(t_i) = 0$ for $i = 1,\ldots,m$ then $(x+\varepsilon a, y+\varepsilon b)$ is an admissible function on $[0,t_m]$ for all sufficiently small $|\varepsilon|$ and hence

$$0 = \int_0^{t_m} [2\ddot{a}\ddot{x} - 3\varkappa^2\dot{a}\dot{x}] = \int_0^{t_m} [2\ddot{b}\ddot{y} - 3\varkappa^2\dot{b}\dot{y}]$$

for all $a, b \in C_0^\infty(t_i, t_{i+1})$, $i = 1,\ldots,m-1$, so that (1.2) holds with $\lambda = 0$. These two equations imply (9.3) with $\lambda = 0$. □

Remark. Note that if for some i there is a point $s \in (t_i, t_{i+1})$ with $x(s) = x(t_i)$ and $y(s) = y(t_i)$ (the case, for example, of a loop), then the functions x and y are C^∞ across the point t_i since $[t_1,\ldots,t_{i-1},s,t_{i+1},\ldots,t_m]$ could have been chosen as our original set of interpolation nodes.

9.2 The Case for $p = \infty$

We now deal with the case of Example 1.2 and consider the properties of solutions to the problem

$$(9.12) \qquad \inf\{\|\theta(x,y)\|_{L^\infty} : (x,y) \in U\}$$

where the (negative) curvature is given by

$$\theta(x,y) = (\ddot{x}\dot{y}-\dot{x}\ddot{y})(\dot{x}^2+\dot{y}^2)^{-5/4}$$

and U consists of all pairs (x,y) in $W^{2,\infty}(0,L) \times W^{2,\infty}(0,L)$ such that the planar curve $t \to (x(t),y(t))$ passes through the specified set $\mathscr{P} = \{(x_i,y_i)\}_{i=1}^m$ and has length L or less. The analysis is much like that of the preceding section but there are distinct differences both in the techniques and the conclusions.

We know from Example 1.2 that (9.12) admits a solution (x,y) parametrized by arc-length so that $\dot{x}^2 + \dot{y}^2 = 1$. Let $\{t_i\}_{i=1}^m$ be chosen so that $x(t_i) = x_i$ and $y(t_i) = y_i$ for $1 \leq i \leq m$. Let u, v be elements of C^∞ which vanish at the points $\{t_i,\ldots,t_m\}$ and for which

$$\int_0^L \dot{u}\dot{x} + \dot{v}\dot{y} < 0.$$

Then we must have

$$(9.13) \qquad \|\theta(x+\epsilon u,y+\epsilon v)\| \geq \alpha$$

for all sufficiently small $\epsilon > 0$. Hence,

$$\|\theta(x,y)+\theta'(u,v)\| \geq \alpha$$

as in Theorem 3.3, where θ' is the Frechet derivative of θ at (x,y) with range in $L^\infty(0,L)$ given by

$$\theta'(u,v) = \ddot{u}\dot{y} - \ddot{y}\dot{u} + \ddot{x}\dot{v} - \dot{x}\ddot{v} - (5/2)\kappa(\dot{x}\dot{u}+\dot{y}\dot{v}) \tag{9.14}$$

and $\kappa = \ddot{x}\dot{y} - \dot{x}\ddot{y}$ is $\theta(x,y)$ since $\dot{x}^2 + \dot{y}^2 \equiv 1$. Let V consist of those elements (u,v) in $W^{2,\infty}(0,L) \times W^{2,\infty}(0,L)$ which vanish at $\{t_1,\ldots,t_m\}$ and for which

$$\int_0^L \dot{x}\dot{u} + \dot{y}\dot{v} \leq 0. \tag{9.15}$$

A continuity argument gives

$$\|\theta(x,y)+\theta'(u,v)\| \geq \alpha, \quad (u,v) \in V. \tag{9.16}$$

Let V_0 be those elements in V for which equality holds in (9.15). V_0 is subspace of $X = W^{2,\infty} \times W^{2,\infty}$. We shall show that the annihilator of $\theta' V_0$ is a subspace of $L^1(0,L)$ of dimension at most $2m - 1$ and shall analyze the structure of the element in this subspace for which $\theta(x,y)$ achieves its norm as a linear functional, as in Theorem 3.3.

Suppose $h \in L^1(0,L)$ and

$$\int_0^L h\theta'(u,v) = 0, \quad (u,v) \in V_0. \tag{9.17}$$

If we let $A = -\ddot{y} - 5/2\, \dot{x}\kappa$, $B = \dot{y}$, $C = \ddot{x} - 5/2\, \dot{y}\kappa$, and $D = -\dot{x}$, then (9.17) may be written

$$\int_0^L h(A\dot{u}+B\ddot{u}+C\dot{v}+D\ddot{v}) = 0, \quad (u,v) \in V_0. \tag{9.18}$$

Taking $v = 0$ and then $u = 0$ we deduce directly (or with a version of Lemma 9.2) that

$$-G + Bh = \lambda x + ct + c_1 \tag{9.19}$$

and

$$-H + Dh = \mu y + dt + d_1 \tag{9.20}$$

on (t_i, t_{i+1}) where $\dot{G} = hA$, $\dot{H} = hC$, and λ, μ, c, d, c_1, d_1 are constants which may depend on i and $\lambda \geq 0$, $\mu \geq 0$. A simple argument much like that at the end of Theorem 9.1 (where it was proved that $\mu_i = \lambda_j$) shows that $\mu = \lambda$. It also follows from (9.19) and (9.20) that both Bh and Dh lie in $W^{1,1}(0,L)$. Since both B and D are in $W^{1,\infty}(0,L)$ and $B^2 + D^2 \equiv 1$ we find that $h \in W^{1,1}(0,L)$. Thus we may differentiate both sides of equation (9.19), (9.20) to get

$$h(\dddot{y}+5/2\varkappa\dot{x}) + \ddot{h}\dot{y} + h\ddot{y} = \lambda\dot{x} + c \tag{9.21}$$

and

$$h(-\dddot{x}+5/2\varkappa\dot{y}) - \ddot{h}\dot{x} - h\ddot{x} = \lambda\dot{y} + d. \tag{9.22}$$

Now multiply (9.21) by $\ddot{x}$ and (9.22) by $\ddot{y}$ and add the resulting equations to get

$$\dot{h}\varkappa = c\ddot{x} + d\ddot{y}. \tag{9.23}$$

Next multiply (9.21) by $\dot{x}$ and (9.22) by $\dot{y}$ and add the resulting equation to get

$$(9.24) \qquad 1/2\, h\kappa = \lambda + c\dot{x} + d\dot{y}.$$

Thus,

$$(h\kappa)^{\cdot} = 2\dot{h}\kappa$$

or

$$(9.25) \qquad \dot{h}\kappa = h\dot{\kappa}.$$

Finally, multiply (9.21) by $\dot{y}$ and (9.22) by $-\dot{x}$ and add the resulting equations to obtain

$$(9.26) \qquad \dot{h} = c\dot{y} - d\dot{x}.$$

Equation (9.26) shows that the L^1-annihilator of $\theta' V_0$ is finite-dimensional. Thus there is an $h \in L^1$ of norm one with h orthogonal to $\theta' V_0$ and

$$\alpha = \int_0^L h\theta(x,y) = \int_0^L h(t)\kappa(t)\,dt.$$

Thus, $h\kappa \geq 0$ a.e. and $|\kappa| = \alpha$ a.e. where $h \neq 0$. Since h is continuous we find that $\kappa = \pm\, \alpha$ on the open intervals in (t_i, t_{i+1}) where $h \neq 0$. But (9.25) then implies that h is constant on these open intervals. Since h is continuous, this implies that h is constant on (t_i, t_{i+1}) so that κ is constant on (t_i, t_{i+1}) if $h \not\equiv 0$. Hence, if (t_i, t_{i+1}) is an interval on which $h \not\equiv 0$, then $t \to (x(t), y(t))$, $t_i \leq t \leq t_{i+1}$, is an arc of a circle. If h is identically zero on some (t_j, t_{j+1}) we can repeat the entire minimization on this interval, using only admissible functions which agree with (x,y) outside of this interval and obtain a solution which is an arc of a circle on (t_j, t_{j+1}). Continue this process until all the intervals $\{(t_i, t_{i+1})\}_{i=1}^{m-1}$ are used. We obtain in this way a solution consisting entirely of arcs of circles. We state these results formally.

Theorem 9.5. Let (x,y) be a solution of (9.12) parametrized by arc length. Then there is some interval (t_i, t_{i+1}) on which the curve $t \to (x(t), y(t))$, $t_i \leq t \leq t_{i+1}$ is an arc of a circle of radius $1/\alpha$. Further, there is a solution of (9.12) which consists entirely of arcs of circles of radius at least $1/\alpha$, with no more than 1 arc for each interval (t_i, t_{i+1}), $i = 1,\ldots,m - 1$. Finally, every solution (x,y) satisfies the equations (9.21), (9.22) and (9.25) locally, where $\varkappa$ is the (negative) curvature, $\varkappa = \ddot{x}\dot{y} - \ddot{y}\dot{x}$ and $\lambda \geq 0$ does not depend on i.

Remarks. The necessity of the condition of bounded length for the existence of an elastica with minimum strain energy was stated in the report [9.2], which predates the paper [9.1]. An example is given in [9.2] of pin support array locations for which no stable equilibrium configuration exists. Also, the significance of the parameter λ remaining constant from one subinterval to the next is related in [9.2] to the equality of certain tension parameters.

It has been known since the time of Euler that the solutions of the equation $\ddot{\varkappa} + \varkappa^3/2 = 0$ involve the elliptic functions. This problem, together with attempts at rectifying the ellipse, led to their systematic development.

Section 9.1 is excerpted from the authors' paper [9.4]. Another topic considered there, which we have not included, is the verification of the continuity of the curve $L \to \alpha(L)$ and the fact that there are cardinality c unstable equilibrium configurations. Here an unstable equilibrium configuration is one whose strain energy α can be strictly decreased for every $\epsilon > 0$, by admitting curves of length $L + \epsilon$.

Sections 9.1 and 9.2, at least in the setting of the minimum curvature problems in L^2 and L^∞, generalize Theorems 3.2 and 3.3. Specifically, in the terminology of those earlier theorems, the image of U_0 under the Frechet derivative of T constitutes a larger set than the admissible perturbations of Tx_0 which fail to decrease α. In the earlier theorems, the entire image, acting as a perturbation set, failed to decrease α. In the present setting, only a proper cone is admissible due to the length restriction and this leads to the Lagrange multiplier results of both sections.

Our derivation of the equations (9.21), (9.22), and (9.25), satisfied locally by the solution (x,y) of the L^∞ problem of minimum curvature, is new. The device used to obtain a solution pieced together by arcs of circles, which consists of restricting attention to a given subinterval (t_i, t_{i+1}), was used by the authors [6.1] in another context, to obtain piecewise perfect (generalized) spline solutions to L^∞ minimization problems on a compact interval. Although the devices do not exactly coincide, they share the common property of selecting a solution which actually solves a sequence of intrinsic minimization problems. We mention, finally, the references [9.5], [9.6] and [9.7], particularly [9.5].

REFERENCES

9.1 G. Birkhoff and C. R. DeBoor, "Piecewise polynomial interpolation and approximation," Approximation of Functions (H. L. Garabedian, editor), Elsevier, New York and Amsterdam, 1965, pp. 164-190.

9.2 G. Birkhoff, H. Burchard and D. Thomas, "Nonlinear interpolation by splines, pseudosplines and elastica," General Motors Research Laboratory Publication 468, Warren, Michigan, 1965.

9.3 F. Deutsch, "Simultaneous interpolation and approximation in linear spaces," SIAM J. Appl. Math. 14 (1966), 1180-1190.

9.4 S. D. Fisher and J. W. Jerome, "Stable and unstable elastica equilibrium and the problem of minimum curvature," J. Math. Anal. Appl., to appear.

9.5 E. H. Lee and G. E. Forsythe, "Variational study of nonlinear spline curves," SIAM Review, 15 (1973), pp. 120-133.

9.6 J. L. Lions and G. Stampacchia, "Variational inequalities," Comm. Pure. Appl. Math., 20 (1967), pp. 493-519.

9.7 A. E. H. Love, The Mathematical Theory of Elasticity (4th ed.), Cambridge Univ. Press, London, 1927.

PART IV. CONVERGENCE THEOREMS

§10. Approximation by Extremals of Nonlinear Differential Expressions in One Variable and Quadratic Forms in Several Variables

10.1 Functions of One Variable

Let φ and ψ be continuous real-valued functions on $[a,b] \times \mathbb{R}^n$. In Example 1.3 and Theorem 3.5 was discussed the minimization in $L^p(a,b)$ of the expression

$$(10.1) \qquad Tf(t) = \varphi(t,f(t),\ldots,f^{(n-1)}(t))D^n f(t) + \psi(t,f(t),\ldots,f^{(n-1)}(t))$$

over a convex subset of $W^{n,p}(a,b)$. For completeness, we shall summarize the basic existence theorem which is a consequence of Theorems 1.1 and 1.2.

Theorem 10.1. Let U be any closed convex subset of $W^{n,p}(a,b)$, $1 < p \leq \infty$, which is weak-* closed if $p = \infty$. Let T be given by (10.1) and suppose there exists a minimizing sequence for the problem

$$(10.2) \qquad \alpha = \inf\{\|Tf\|_p : f \in U\}$$

which is bounded in $W^{n,p}(a,b)$. Then the extremal problem (10.2) has a solution $s \in U$.

Proof: It clearly suffices to prove that T maps weakly (resp. weak-*) convergent sequences onto weakly (resp. weak-*) convergent sequences if $1 < p < \infty$ (resp. $p = \infty$). However, as remarked in Example 1.3, weak (or weak-*) convergence of f_ν implies uniform convergence of derivatives through order $n - 1$; it then follows that $D^n f_\nu$ is weakly (weak-*) convergent in $L^p(a,b)$ and $\varphi(\cdot,f,\ldots,f^{(n-1)})$, $\psi(\cdot,f,\ldots,f^{(n-1)})$ are uniformly convergent. It follows that Tf_ν is weakly (weak-*) convergent in $L^p(a,b)$ and the result is a consequence of Theorem 1.1 for $1 < p < \infty$ and Theorem 1.2 for $p = \infty$.

Now let Δ be a partition of $[a,b]$ given by

$$a \leq x_1 < \ldots < x_m \leq b.$$

We define,

$$\overline{\Delta} = \max\{x_1-a, x_2-x_1, \ldots, b-x_m\}.$$

We have

<u>Theorem 10.2</u>. Suppose that Δ is a partition of $[a,b]$ consisting of at least $n + 1$ points and let U be the flat in $W^{n,p}(a,b)$, $1 \leq p \leq \infty$, defined by

$$U = \{u : u(x_i) = f(x_i),\ i = 1,\ldots,m\}$$

for a fixed function $f \in W^{n,p}(a,b)$. If (10.2) has a solution $s \in U$ and if there is a nonnegative monotone increasing function σ on $[0,\infty)$ such that

$$(10.3) \qquad \|D^n(u-s)\|_p \leq \sigma(\|Tu\|_p),\ \forall\ u \in U,$$

then we have the estimates, for each $0 \leq j < n$ and $1 \leq r \leq \infty$,

$$\text{(10.4)}\quad \begin{aligned} &\text{(i)}\quad \|D^j(f-s)\|_r \le C_j(\overline{\Delta})^{n-j-1/p+1/r}\sigma(\|Tf\|_p),\ (r \ge p) \\ &\text{(ii)}\quad \|D^j(f-s)\|_r \le \overline{C}_j(b-a)^{1/r-1/p}(\overline{\Delta})^{n-j}\sigma(\|Tf\|_p), \\ &\qquad\qquad (1 \le r \le p). \end{aligned}$$

Here, the constants are given by

$$C_j = \frac{n!}{j!n^{1/p-1/r}},\ \overline{C}_j = \frac{n!}{j!}.$$

<u>Proof</u>. Suppose first that $p < \infty$. (10.4ii) for $1 \le r < p$ follows from the inequality for $r = p$ and from Hölder's inequality. Therefore, assume $r \ge p$. Select a subset $\Delta^{(0)} : \xi_0^{(0)} < \ldots < \xi_{N_0}^{(0)}$ of Δ containing at least $n + 1$ points such that $\overline{\Delta}^{(0)} = \overline{\Delta}$. Since $(f-s) \in C^{n-1}[a,b]$, there exist points $\Delta^{(j)} : \xi_0^{(j)} < \ldots < \xi_{N_j}^{(j)}$ in $[a,b]$, by Rolle's theorem, such that $0 = D^jf(\xi_\ell^{(j)}) - D^js(\xi_\ell^{(j)})$ for each $j = 1,\ldots,n-1$ with $\overline{\Delta}^{(j)} \le (j+1)\overline{\Delta}$. We establish now the two inequalities

$$\text{(10.5)}\quad \|D^{n-1}f-D^{n-1}s\|_r \le n^{1+1/r-1/p}\|D^nf-D^ns\|_p\overline{\Delta}^{1+1/r-1/p},$$

$$\text{(10.6)}\quad \|D^jf-D^js\|_r \le (j+1)\|D^{j+1}f-D^{j+1}s\|_r\overline{\Delta},\ 0 \le j \le n-2.$$

It is clear that (10.5) and (10.6) imply, for $0 \le j < n$,

$$\text{(10.7)}\quad \|D^jf-D^js\|_r \le \frac{n!}{j!n^{1/p-1/r}}\|D^nf-D^ns\|_p\overline{\Delta}^{n-j+1/r-1/p}$$

and (10.4) follows from (10.7) upon applying the inequality

$$\|D^nf-D^ns\|_p \le \sigma(\|Ts\|_p).$$

Now (10.5) and (10.6) easily follow from the more general inequality, valid for $r \geq \nu \geq p$ whenever $\|D^j f - D^j s\|_\nu$ exists,

$$(10.8)\quad \|D^j f - D^j s\|_r \leq (j+1)^{1+1/r-1/\nu} \|D^{j+1} f - D^{j+1} s\|_\nu (\overline{\Delta})^{1+1/r-1/\nu}.$$

Now, if $\xi_\ell^{(j)} \leq t \leq \xi_{\ell+1}^{(j)}$ for $0 \leq j \leq n-1$ and $0 \leq \ell \leq N_j - 1$, we have

$$(10.9)\quad D^j f(t) - D^j s(t) = \int_{\xi_\ell^{(j)}}^{t} [D^{j+1} f(x) - D^{j+1} s(x)]\,dx.$$

If $r < \infty$, by Hölder's inequality we have

$$(10.10)\quad \{\int_{\xi_\ell^{(j)}}^{\xi_{\ell+1}^{(j)}} |D^j f(t) - D^j s(t)|^r\}^{1/r}$$

$$\leq ((j+1)\overline{\Delta})^{1-1/\nu+1/r} \times \{\int_{\xi_\ell^{(j)}}^{\xi_{\ell+1}^{(j)}} |D^{j+1} f(x) - D^{j+1} s(x)|^\nu dx\}^{1/\nu}.$$

Similar inequalities hold on the intervals $[a, \xi_0^{(j)}]$ and $[\xi_{N_j}^{(j)}, b]$. Recalling the elementary inequality, for $\mu_\ell \geq 0$,

$$(10.11)\quad \{\sum_{\ell=0}^{N} \mu_\ell^r\}^{1/r} \leq \{\sum_{\ell=0}^{N} \mu_\ell^\nu\}^{1/\nu}, \quad r \geq \nu \geq 1,$$

we have from (10.11), upon setting $\mu_{\ell+1}$ equal to the left-hand side of (10.10) for $0 \leq \ell \leq N_j - 1$, and setting μ_0 and μ_{N_j+1} equal respectively to the corresponding contributions to the L^r-norm of $D^j(f-s)$ on $[a, \xi_0^{(j)}]$ and $[\xi_{N_j}^{(j)}, b]$,

$$\|D^j f - D^j s\|_r \leq \{\sum_{\ell=0}^{N_j+1} \mu_\ell^r\}^{1/r} \leq \{\sum_{\ell=0}^{N_j+1} \mu_\ell^\nu\}^{1/\nu}$$

$$\leq ((j+1)\overline{\Delta})^{1-1/\nu+1/r} \|D^{j+1} f - D^{j+1} s\|_\nu.$$

This establishes (10.8) for $r < \infty$.

If $r = \infty$, select ℓ and $t \in [\xi_\ell^{(j)}, \xi_{\ell+1}^{(j)}]$ (or $t \in [a, \xi_0^{(j)}]$, $[\xi_{N_j}^{(j)}, b]$) such that

$$\|D^j f - D^j s\|_\infty = |D^j f(t) - D^j s(t)|, \quad 0 \leq j \leq n - 1.$$

From (10.9) and Hölder's inequality we have

$$\|D^j f - D^j s\|_\infty \leq ((j+1)\overline{\Delta})^{1-1/\nu} \|D^{j+1} f - D^{j+1} s\|_\nu ,$$

where we permit $\nu = \infty$. This establishes (10.8) for $r = \infty$, which completes the proof when $p < \infty$. Suppose now that $p = \infty$.

Suppose first that $r = \infty$. For fixed j, $0 \leq j \leq n - 1$, choose $t \in [a,b]$ such that

$$\|D^j(f-s)\|_\infty = |D^j(f-s)(t)|. \tag{10.12}$$

Now select $\xi_\ell^{(j)}$, $0 \leq \ell \leq N_j$, such that $|t - \xi_\ell^{(j)}| \leq (j+1)\overline{\Delta}$. By (10.9) and (10.12), we have

$$\|D^j(f-s)\|_\infty \leq (j+1)\overline{\Delta}\|D^{j+1}(f-s)\|_\infty . \tag{10.13}$$

By induction, we obtain from (10.13),

$$\|D^j(f-s)\|_\infty \leq (n!/j!)(\overline{\Delta})^{n-j}\|D^n(f-s)\|_\infty \tag{10.14}$$

and (10.4), for $r = \infty$, follows from (10.14) and the inequality

$$\|D^n(f-s)\|_\infty \leq \sigma(\|Tf\|_\infty). \tag{10.15}$$

The inequality (10.4), for $1 \leq r < \infty$, follows from the obvious inequality

$$(10.16) \qquad \|D^j(f-s)\|_r \leq (b-a)^{1/r}\|D^j(f-s)\|_\infty$$

and from the established inequality (10.4) for $r = \infty$. This completes the proof of Theorem 10.2. □

The remarkable fact concerning solutions of (10.2) is that simple hypotheses of smoothness of φ and ψ guarantee the existence of a linear function σ satisfying the hypotheses of Theorem 10.2 and thus the estimates (10.4) hold in this case. Specifically, we assume

$$(10.17) \quad \varphi(t,\xi_0,\ldots,\xi_{n-1}) > 0, \text{ for all } (t,\xi_0,\ldots,\xi_{n-1}) \in [a,b] \times \mathbb{R}^n$$

$$(10.18) \quad \zeta(t,\xi_0,\ldots,\xi_{n-1}) = \psi(t,\xi_0,\ldots,\xi_{n-1})/\varphi(t,\xi_0,\ldots,\xi_{n-1})$$

is differentiable on $\mathbb{R}^n$ for each $t \in [a,b]$ with each first order partial derivative continuous on $[a,b] \times \mathbb{R}^n$.

<u>Theorem 10.3</u>. Suppose that U is defined by interpolation, at the points of a partition Δ, of a function $f \in W^{n,p}(a,b)$, that (10.17) and (10.18) hold and that (10.2) has a solution. If $\bar{\Delta}$ satisfies

$$(10.19) \qquad n!\left(\frac{b-a}{n}\right)^{1/p} \sum_{j=0}^{n-1} \frac{K_j}{j!} \bar{\Delta}^{n-j-1/p} \leq \rho, \; 0 \leq \rho < 1,$$

$$K_j = \sup_{\substack{u \in U \\ a \leq t \leq b}} \left| \frac{\partial}{\partial \xi_j} \zeta(t,\xi_0,\ldots,\xi_{n-1})_{\xi_\nu = u^{(\nu)}(t)} \right| < \infty,$$

$$j = 0,\ldots,n-1$$

then the estimates (10.4) hold with

$$\sigma(t) = (2M/(1-\rho))t, \ t \in [0,\infty) \tag{10.20}$$

$$M = \sup_{\substack{a \le t \le b \\ u \in U}} |\varphi(t,u(t),\ldots,u^{(n-1)}(t))|^{-1}$$

Proof. We need only show that the function σ of (10.20) satisfies

$$\|D^n u - D^n s\|_p \le \sigma(\|Tu\|_p) \tag{10.21}$$

for all $u \in U$ provided $\overline{\Delta}$ satisfies (10.19). However,

$$\begin{aligned} \|D^n u - D^n s\|_p \le\ & \|Tu/\varphi(\cdot,u(\cdot),\ldots,u^{(n-1)}(\cdot))\|_p \\ & + \|Ts/\varphi(\cdot,s(\cdot),\ldots,s^{(n-1)}(\cdot))\|_p \\ & + \|\zeta(\cdot,u(\cdot),\ldots,u^{(n-1)}(\cdot)) \\ & \quad - \zeta(\cdot,s(\cdot),\ldots,s^{(n-1)}(\cdot))\|_p. \end{aligned} \tag{10.22}$$

By the differentiability assumption on ζ, we have, for each $t \in [a,b]$,

$$\begin{aligned} & |\zeta(t,u(t),\ldots,u^{(n-1)}(t)) - \zeta(t,s(t),\ldots,s^{(n-1)}(t))| \\ & \le \sum_{j=0}^{n-1} K_j |u^{(j)}(t) - s^{(j)}(t)| \le \sum_{j=0}^{n-1} K_j \|u^{(j)} - s^{(j)}\|_\infty, \end{aligned} \tag{10.23}$$

where K_j is defined in (10.19). Now from (10.7),

$$\|D^j u - D^j s\|_\infty \le \frac{n!}{j!\,n^{1/p}} \overline{\Delta}^{n-j-1/p} \|D^n f - D^n s\|_p \tag{10.24}$$

so that (10.21) follows from (10.19), (10.22), (10.23) and (10.24). □

Remark. From the proof of Theorem 10.2, it is clear that the constants C_j and $\overline{C}_j$ can be improved slightly if, for every $x \in \Delta$, $D^\nu f(x) = D^\nu s(x)$ for $0 \leq \nu \leq j_0$, for some $0 \leq j_0 \leq n-1$. Indeed, the more general constants are given by

$$C_{j,j_0} = \frac{(n-j_0)!}{(j-j_0)!(n-j_0)^{1/p-1/r}}, \; \overline{C}_{j,j_0} = \frac{(n-j_0)!}{(j-j_0)!}, \; j \geq j_0;$$

$$C_{j,j_0} = \frac{(n-j_0)!}{(n-j_0)^{1/p-1/r}}, \qquad \overline{C}_{j,j_0} = (n-j_0)!, \; 0 \leq j < j_0.$$

Also, the constants K_j of (10.19) are clearly finite if $U \subset W^{n-1,\infty}(a,b)$ or if the partial derivatives $\frac{\partial}{\partial \xi_0}, \ldots, \frac{\partial}{\partial \xi_{n-1}}$ are bounded. When T is a nonsingular linear differential operator of the form

$$(10.25) \qquad Tf(t) = D^n f(t) + a_{n-1}(t) D^{n-1} f(t) + \ldots + a_0(t),$$

in which case $\partial \zeta(t, \xi_0, \ldots, \xi_{n-1})/\partial \xi_j = a_j(t)$, $a \leq t \leq b$, $0 \leq j \leq n-1$, then $K_j = \sup_{a \leq t \leq b} |a_j(t)|$.

In the following theorem we shall choose T to be the linear operator given by (10.25) and we shall use the notion of k-width to obtain lower bounds for approximation. Specifically, let

$$(10.26) \qquad \mathscr{B} = \{f \in W^{n,p} : \|Tf\|_p \leq 1\}.$$

The k-widths of the class $\mathscr{B}$ in $L^p(a,b)$ for $1 < p < \infty$ and in $C[a,b]$ for $p = \infty$ are known (cf. [10.3 and [10.5], Th. 1, p. 135]) and are given asymptotically by

$$(10.27) \qquad C_1 k^{-n} \leq d_k(\mathscr{B}) \leq C_2 k^{-n}, \; k \to \infty,$$

where $0 < C_1 \leq C_2$. This means, in particular that for any sequence M_k of subspaces of $L^p(a,b)$, for $1 < p < \infty$, and of $C[a,b]$, for $p = \infty$, of dimension k there is a corresponding sequence of members $f_k \in \mathscr{B}$ such that

(10.28) $$\|f_k - f_k^*\|_p \geq C_1 k^{-n}$$

for sufficiently large k, where $f_k^* \in M_k$ is an element of best approximation to f_k. We may now state

Theorem 10.4. Let T be given by (10.25) and for $k \geq 1$ consider the partitions $\Delta : a = x_1 < \ldots < x_k = b$ of $[a,b]$ satisfying $\overline{\Delta}/\underline{\Delta} \leq C$ for $C > 0$ where

$$\underline{\Delta} = \min_{1 \leq i \leq k-1} \{x_{i+1} - x_i\}.$$

Then the approximation procedure which, for each $f \in \mathscr{B}$, with $\mathscr{B}$ defined by (10.26), selects the extremal solution s_k of

(10.29) $$\|Ts_k\|_p = \min\{\|Tu\|_p : u(x_\nu) = f(x_\nu),\ \nu = 1,\ldots,k\}$$

is optimal in k with respect to any sequence of projections onto subspaces M_k of $L^p(a,b)$, $1 < p < \infty$, and $C[a,b]$, $p = \infty$, of dimension k. Specifically, the estimates

(10.30) $$\|f - s_k\|_p \leq Ck^{-n},\ k = 1,2,\ldots$$

hold, for $C > 0$ independent of f and k, and, moreover, there is a constant $C_1 > 0$ such that, for any sequence M_k of linear subspaces of $L^p(a,b)$, $1 < p < \infty$, and $C[a,b]$, $p = \infty$, there are corresponding sequences $\{f_k\} \subset \mathscr{B}$ and $\{f_k^*\}$, f_k^* a best approximation in M_k to f_k

such that (10.28) holds for sufficiently large k.

Proof. The existence of solutions of (10.29) follows from Example 2.1. (10.30) is a consequence of Theorem 10.3 and the simple inequality

$$\overline{\Delta} \leq C\underline{\Delta} \leq C\,\frac{b-a}{k-1}\ ,\ k \geq 2.$$

The remaining assertions follow from the k-width results. □

We close this section with an application of Theorem 3.2 to the characterization of solutions of (10.2) where the slab U is defined by inequality constraints.

Theorem 10.5. Let T be given by (10.1), where φ satisfies (10.17) and where φ and ψ are continuous on $[a,b] \times \mathbb{R}^n$ and differentiable on $\mathbb{R}^n$ for each $t \in [a,b]$, with each first order partial derivative continuous on $[a,b] \times \mathbb{R}^n$. Let the slab U of (10.2) be defined by the extended Hermite-Birkhoff inequality constraints of (3.18ii). Then, every solution s of (10.2) for $1 < p < \infty$ satisfies locally the nonlinear differential equation

$$(10.31)\qquad (T')^*|Ts|^{p-1}\ \text{signum}\ Ts = 0 \text{ on } (x_i, x_{i+1})$$

and, moreover, (3.18iii), (3.18iv) and (3.18v) hold if, in the application of the functionals R_{ij}, is understood the replacement by T of L. Here T' is the Frechet derivative of T at s given by

$$T'f = \sum_{i=0}^{n-1} \{D^n s D^i f \left.\frac{\partial\varphi}{\partial\xi_i}\right|_{\xi_j = D^j s} + D^i f \left.\frac{\partial\psi}{\partial\xi_i}\right|_{\xi_j = D^j s}\}$$

$$+ \varphi(\cdot, s(\cdot), \ldots, D^{n-1}s(\cdot))D^n f$$

and $(T')^*$ is the formal adjoint of T'.

<u>Proof</u>. (10.31) follows from Theorem 3.2 (note that T' satisfies Theorem 2.3 so that $T'U_0$ is closed) and from the theory of distributions [3.2, Ch. 8]. The remaining conclusions follow from the proof of Theorem 3.7 and the orthogonality relation (3.2i),

$$\int_a^b T'f|Ts|^{p-1} \text{ signum } Ts = 0, \text{ for all } f \in U_0.$$

Here U_0 is given in Theorem 3.2. □

10.2 Functions of Several Variables

In this section, we consider the minimization of quadratic functionals which satisfy certain coerciveness relationships in Sobolev spaces. Our notion of coerciveness is somewhat more general than that usually employed and its formulation is in fact the content of Gårdings's inequality. Although our applications are primarily to functions of several variables, they apply equally well to functions of one variable and thus cover situations in which the functional is not directly of the form $\|Tf\|$.

Let H and V be Hilbert spaces, $V \subseteq H$, such that

$$\text{(10.32)} \qquad \begin{aligned} &\text{(i)} \quad I : V \to H \text{ is compact} \\ &\text{(ii)} \quad V \text{ is dense in } H. \end{aligned}$$

Let W be a Hilbert space and Γ a linear mapping of V into W with kernel V_0 such that

$$\text{(10.33)} \qquad \begin{aligned} &\text{(i)} \quad \Gamma \text{ is continuous and surjective, and,} \\ &\text{(ii)} \quad V_0 \text{ is dense in } H. \end{aligned}$$

Finally, let $B(u,v)$ be a bilinear form on V satisfying

(10.34)

(i) $B(u,v)$ is continuous on V, i.e.
$|B(u,v)| \leq C\|u\|\ \|v\|$ for all u, $v \in V$.

(ii) $B(u,u)$ is coercive, i.e., there exist positive constants C and λ such that

$$B(u,u) + C(u,u)_H \geq \lambda(u,u)_V \text{ for all } u \in V.$$

The following consequence of the Riesz-Fredholm-Schauder theory is taken from Aubin [10.1, Theorem 2.3, p. 181].

Theorem 10.6. Suppose that (10.32), (10.33) and (10.34) are satisfied, that ℓ is a continuous linear functional on V_0 and that $u_0 \in V_0$ and

(10.35) $B(u_0,u) = 0$ for all $u \in V_0$ implies that $u_0 = 0$.

Then there exists a unique element $u_0 \in V_0$ satisfying,

$$B(u_0,u) = \ell(u), \text{ for all } u \in V_0. \tag{10.36}$$

We proceed immediately to the applications. For the major application we consider any bounded open subset Ω of $\mathbb{R}^n$. With $V = W_0^m(\Omega)$ and $H = L^2(\Omega)$, (10.32) is satisfied [1.1, p. 99]. Now let $B(u,v)$ be given by

$$B(u,v) = \int_\Omega \sum_{|\alpha|,|\beta| \leq m} c_{\alpha\beta} D^\alpha u D^\beta v \tag{10.37}$$

on $W^m(\Omega)$ where $c_{\alpha\beta} \in L^\infty(\Omega)$ for $|\alpha|$, $|\beta| \leq m$ and $c_{\alpha\beta} \in C(\overline{\Omega})$ for $|\alpha| = |\beta| = m$. Then (10.34i) holds by the definition of $B(u,v)$ and

(10.34ii) is a consequence of Gårding's inequality [1.1, p. 78] provided $B(u,u)$ is uniformly strongly elliptic in Ω:

$$\Sigma_{|\alpha|+|\beta|=m}\, c_{\alpha\beta}(x)\xi^{\alpha+\beta} \geq E_0|\xi|^{2m}, \quad \xi \in \mathbb{E}^n \tag{10.38}$$

for all $x \in \Omega$ and some positive constant E_0.

We observe that, if $U \in W^m(\Omega)$, then a solution $U_0 \in W^m(\Omega)$ of the generalized Dirichlet problem: $U - U_0 \in W_0^m(\Omega)$, and

$$B(U_0,u) = 0, \text{ for all } u \in W_0^m(\Omega) \tag{10.39}$$

exists as a consequence of Theorem 10.6, if (10.35) holds. Here we must take $W = \{0\}$, $V = W_0^m(\Omega)$, $\Gamma = 0$ and $\ell(u) = -B(U,u)$. Now let $W = \mathbb{R}^N$ and let $F_1,\ldots,F_N$ be N non-zero linearly independent bounded linear functionals on $W_0^m(\Omega)$. In the notation of (10.33), we define

$$\Gamma u = (F_1u,\ldots,F_Nu), \; u \in W_0^m(\Omega). \tag{10.40}$$

Notice that (10.33i) holds. We explicitly assume (10.33ii); for example, this holds if $m > n/2$ and

$$F_iu = u(x_i), \; i = 1,\ldots,N \tag{10.41}$$

for distinct points $x_1,\ldots,x_N \in \Omega$.

<u>Theorem 10.7</u>. Let $B(u,v)$ be the bilinear form defined by (10.37) and satisfying the property that if $u_0 \in W_0^m(\Omega)$ and if $B(u_0,u) = 0$ for all $u \in W_0^m(\Omega)$, then $u_0 = 0$. Suppose also that $B(u,u)$ is coercive and that the kernel V_0 of Γ, with Γ defined by (10.40), satisfies (10.33ii). Then, for each $r \in \mathbb{R}^N$, there exists a unique

$v \in W_0^m(\Omega)$ satisfying $\Gamma v = r$ and

$$B(v,u) = 0, \text{ for all } u \in V_0. \tag{10.42}$$

In particular, if $m > n/2$ and the functionals F_i are given by (10.41), then for every $U \in W^m(\Omega)$ there is a unique function $U_0 \in W^m(\Omega)$ such that $U - U_0 \in V_0$ and

$$B(U_0,u) = 0, \text{ for all } u \in V_0. \tag{10.43}$$

Proof. The existence of a unique $v = u_0 \in W_0^m(\Omega)$ satisfying (10.36) is guaranteed by Theorem 10.6 with

$$\ell(u) = -B(Mr,u),$$

where M is a continuous right inverse for Γ. Such an inverse exists by the open mapping theorem. Then, by Theorem 10.6,

$$B(u_0+Mr,u) = 0, \text{ for all } u \in V_0.$$

(10.42) follows with $v = u_0 + Mr$. Now if $m > n/2$ and the F_i are given by (10.41), then we define

$$r = (U(x_1)-\tilde{U}_0(x_1),\ldots,U(x_N)-\tilde{U}_0(x_N))$$

for $U \in W^m(\Omega)$ and $\tilde{U}_0$ the unique solution of the generalized Dirichlet problem:

(i) $U - \tilde{U}_0 \in W_0^m(\Omega)$

(ii) $B(\tilde{U}_0,u) = 0$, for all $u \in W_0^m(\Omega)$.

By the preceding, there is a unique $v \in W_0^m(\Omega)$ satisfying $\Gamma v = r$ and (10.42). We now define $U_0 = \tilde{U}_0 + v$ and (10.43) follows from (10.42) and (10.39). $\square$

<u>Corollary 10.8</u>. If $B(u,v)$ is symmetric on $W_0^m(\Omega)$ and

$$B(u,u) \geq 0, \text{ for all } u \in W_0^m(\Omega),$$

then, under the hypotheses of Theorem 10.7, for each $r \in \mathbb{R}^N$ there exists a unique $v \in W_0^m(\Omega)$ satisfying

$$(10.44) \qquad B(v,v) = \inf\{B(u,u) : u \in W_0^m(\Omega) \text{ and } \Gamma u = r\}.$$

<u>Proof</u>. Let $u \in W_0^m(\Omega)$ satisfy $\Gamma u = r$. Then, if v satisfies (10.42), with $u - v \in V_0$, we have

$$\begin{aligned} 0 \leq B(u-v,u-v) &= B(u,u) - B(u,v) \\ &= B(u,u) - B(v,u) \\ &= B(u,u) - B(v,v). \end{aligned}$$

The uniqueness follows since (10.44) implies (10.42). $\square$

<u>Corollary 10.9</u>. Let the hypotheses of Corollary (10.8) be satisfied, suppose that $m > n/2$ and let the F_i be given by (10.41). For each $U \in W^m(\Omega)$ there exists a unique $U_0 \in W^m(\Omega)$ with $U - U_0 \in V_0$ satisfying

$$B(U_0,U_0) = \inf\{B(u,u) : u \in W^m(\Omega), U-u \in V_0\}.$$

<u>Proof</u>. Let U_0 be given by the second part of Theorem 10.7. Then,

if $u - U_0 \in V_0$, we have

$$0 \leq B(u-U_0,u-U_0) = B(u,u) - B(U_0,U_0). \quad \square$$

<u>Corollary 10.10</u>. Let the coefficients $c_{\alpha\beta}$ of $B(u,v)$ satisfy $c_{\alpha\beta} \in C^{|\beta|+k}(\Omega) \cap C^m(\Omega)$ for some integer $k \geq 0$ with $2m + k > [n/2]$ where $B(u,v)$ satisfies (10.34ii). If the F_i are given by (10.41) with $m > n/2$ and, if (10.35) holds, then the unique function $v \in W_0^m(\Omega)$ which satisfies the variational condition

$$\text{(10.45)} \qquad \begin{aligned} &\text{(i)} \quad B(v,u) = 0, \text{ for all } u \in V_0, \\ &\text{(ii)} \quad \Gamma v = r, \; r \in \mathbb{R}^N, \end{aligned}$$

satisfies the smoothness property:

$$v \in C^h(\Omega'), \text{ where } h = 2m + k - [n/2] - 1, \text{ and}$$

$$\text{where } \Omega' = \Omega - \{x_i\}_{i=1}^N.$$

In particular, if $k > n/2$ then $Lv(x) = 0$ on Ω' where

$$L = \Sigma_{|\alpha|,|\beta| \leq m} (-1)^{|\beta|} D^\beta c_{\alpha\beta} D^\alpha$$

and the derivatives are taken in the classical sense. Moreover, v is C^∞ [real analytic] on Ω' if the $c_{\alpha\beta}$ are.

<u>Proof</u>. See Friedman [4.2, Thm. 16.2, p. 56; Thm. 1.2, p. 205]. The operator L^* is defined by

$$L^*u = \sum_{|\alpha|,|\beta| \leq m} (-1)^{|\alpha|} D^\alpha (c_{\alpha\beta} D^\beta u), \; u \in W^{2m}(\Omega).$$

L is strongly elliptic by (10.34ii) [1.1, Thm. 7.12] and, by (10.45i), v is a weak solution of $Lv = 0$. The assumptions on $c_{\alpha\beta}$ imply the conclusion. □

Corollary 10.11. Suppose that $B(u,v)$ is given by

$$B(u,v) = (Mu, Mv)_{L^2(\Omega)}$$

where

$$Mu = \sum_{|\alpha| \leq m} c_\alpha D^\alpha u$$

with $c_\alpha \in C(\overline{\Omega})$ for $|\alpha| = m$ and $c_\alpha \in L^\infty(\Omega)$ for $|\alpha| < m$. Then, if $B(u,u)$ is coercive and (10.33ii) and (10.35) are satisfied, there exists a unique solution of the minimization problem (10.44). If the c_α satisfy the additional smoothness properties

$$c_\alpha \in C^{m+k}(\Omega) \text{ for some integer } k \geq 0$$

and if the F_i are given by (10.41) with $m > n/2$, then $v \in C^h(\Omega')$ where $h = 2m + k - [n/2] - 1$ and satisfies classically

$$M^*Mv = 0 \text{ on } \Omega' = \Omega - \{x_i\}$$

if $k > n/2$.

Proof. A direct consequence of the two previous corollaries. □

Example 10.1. Let $n = 2$, $\Omega = \{(x,y) : x^2 + y^2 < 1\}$ and let $M = \Delta$, the Laplacian. Then there is a unique u_0 with $u_0(0) = 1$, $u_0 = 0$ on $\{(x,y) : x^2 + y^2 = 1\}$ and $\int_\Omega (\Delta u_0)^2 dx\, dy$ minimal. u_0 must be analytic on $\Omega - \{0\}$ and satisfy $\Delta^2 u_0 = 0$ on $\Omega - \{0\}$.

We complete our discussion of this application by a convergence result. Let $h > 0$ and let

$$G_h = \{h(j_1,\dots,j_n) : j_i \text{ integral}\} \cap \Omega$$

be those points in Ω whose coordinates are integral multiples of h.

Lemma 10.12. There exist positive constants C, C', C'', such that for all h sufficiently small

$$\text{(10.46)} \quad \text{(i)} \quad \|u\|_m^2 \leq 2/\lambda[Ch^n \sum_{x \in G_h} [u(x)]^2 + B(u,u)] \leq C'\|u\|_m^2,$$

$$\text{(ii)} \quad \|u\|_{L^2}^2 \leq C''[h^n \sum_{x \in G_h} [u(x)]^2 + h^{2m}\|u\|_m^2],$$

$$\text{(iii)} \quad h^n \sum_{x \in G_h} [u(x)]^2 \leq C''[\|u\|_{L^2}^2 + h^{2m}\|u\|_m^2]$$

for all $u \in W_0^m(\Omega)$ where B satisfies (10.34) and λ is as in (10.34ii).

Proof. We make use of the complex-valued discrete Fourier transform,

$$\text{(10.47)} \qquad \tilde{u}(\xi) = h^n \sum_{x \in G_h} u(x)e^{-i\langle x,\xi\rangle}$$

if $u \in W_0^m(\Omega)$ and $\langle x,\xi\rangle = \Sigma_{i=1}^n x_i\xi_i$, for $x = (x_1,\dots,x_n)$ and $\xi = (\xi_1,\dots,\xi_n)$. The relation

$$\text{(10.48)} \qquad h^n \sum_{x \in G_h} [u(x)]^2 = (1/2\pi)^n \int_{K_h} |\tilde{u}(\xi)|^2 d\xi$$

holds for the hypercube

$$k_h = \{x : |x_i| \leq \pi/h\}$$

[cf. [10.2, eq. (4.7)] for $u \in W_0^m(\Omega)$. Now it is proved in [10.2; Thm. 5] that the estimate

$$(10.49) \qquad \|\chi_h \tilde{u} - \hat{u}\|_{L^2(\mathbb{R}^n)} \leq C_1 h^m |u|_m$$

holds, where the norms now involve complex-valued functions. χ_h is the characteristic function of k_h and $\hat{u}$ is the Fourier transform of u, defined by,

$$\hat{f}(\xi) = \int_{\mathbb{R}^n} f(x) e^{-i\langle x,\xi\rangle} dx.$$

(10.46ii) is a consequence of (10.48), (10.49) and the Fourier-Plancherel theorem. Indeed,

$$(2\pi)^n \|u\|^2_{L^2(\Omega)} = \|\hat{u}\|^2_{L^2(\mathbb{R}^n)} \leq 2\|\chi_h \tilde{u} - \hat{u}\|^2_{L^2(\mathbb{R}^n)} + 2\|\tilde{u}\|^2_{L^2(k_h)}$$

$$\leq 2(2\pi)^n h^n \sum_{x \in G_h} [u(x)]^2 + 2C_1^2 h^{2m} |u|^2_m$$

which implies (10.46ii). (10.46i) depends upon (10.34ii) and (10.46ii). For,

$$\|u\|^2_m \leq (1/\lambda)[B(u,u) + C\|u\|^2_{L^2}]$$

$$\leq (1/\lambda)[B(u,u) + CC'' h^n \sum_{x \in G_h} [u(x)]^2 + CC'' h^{2m} \|u\|^2_m]$$

and, if $(CC''/\lambda) h^{2m} \leq \lambda' < 1/2$ we have

$$\|u\|^2_m \leq (1/(\lambda(1-\lambda')))[B(u,u) + CC'' h^n \sum_{x \in G_h} [u(x)]^2]$$

from which the first inequality of (10.46i) follows.

To deduce (10.46iii), use (10.48) and obtain

$$(2\pi h)^n \sum_{x \in G_h} [u(x)]^2 = \|\chi_h \tilde{u}\|^2_{L^2(\mathbb{R}^n)}$$

and, by (10.49) and the Fourier-Plancherel theorem we have

$$\|\chi_h \tilde{u}\|^2_{L^2(\mathbb{R}^n)} \leq 2\|\chi_h \tilde{u} - \hat{u}\|^2_{L^2(\mathbb{R}^n)} + 2(2\pi)^n \|u\|^2_{L^2(\Omega)}$$

$$\leq 2C_1^2 h^{2m} |u|^2_m + 2(2\pi)^n \|u\|^2_{L^2(\Omega)}$$

from which (10.46iii) follows. The second inequality of (10.46i) is now immediate. □

Theorem 10.13. Let $B(u,v)$ be given by (10.37) and satisfy the hypotheses that $B(u,v)$ is symmetric and

(10.50) $$\|u\|^2_m \leq C_1 B(u,u) \text{ for all } u \in W^m_0(\Omega).$$

Let the F_i be given by (10.41) where $m > n/2$ and the points $\{x_i\}$ are the points of G_h. Let $U \in W^m_0(\Omega)$ and let v_h be the unique solution of (10.44) where $r = \Gamma U$. Then we have the estimate

(10.51) $$\|U - v_h\|_j \leq C h^{m-j} \|U\|_m, \quad 0 \leq j < m$$

where C is independent of h and U. More generally, if $B(u,v)$ is given by (10.37) and is assumed only to be coercive (condition (10.34ii)) then there exists $h_0 > 0$ such that, for all $0 < h \leq h_0$, (10.44) has a unique solution v_h satisfying the estimate (10.51).

Proof. By [10.2, Thm. 9], there exists an operator

$$S : W_0^m(\Omega) \to W^{m-1}(\Omega)$$

such that Su depends only on the values of u at the points of G_h and such that

$$(10.52) \qquad \|Su-u\|_j \leq C'h^{m-j}\|u\|_m$$

for C' independent of h and u. Su is a piecewise polynomial of degree m - 1 in each variable with partial derivatives continuous through order m - 2. Now suppose that (10.50) holds. Then, $SU = Sv_h$ for each $h > 0$ and, by (10.52) and (10.50) we have,

$$\begin{aligned}\|U-v_h\|_j &\leq \|U-SU\|_j + \|Sv_h-v_h\|_j \\ &\leq C'h^{m-j}[\|U\|_m+\|v_h\|_m] \\ &\leq C'(1+C_1)h^{m-j}\|U\|_m.\end{aligned}$$

If (10.50) fails to hold we make use of Lemma 10.12. Thus, for h sufficiently small

$$|||u||| = \{Ch^n \sum_{x \in G_h} [u(x)]^2 + B(u,u)\}^{1/2}$$

is a norm on $W_0^m(\Omega)$ equivalent to the Sobolev norm. Given $U \in W_0^m(\Omega)$, there clearly exists a unique $v_h \in W_0^m(\Omega)$, of minimal $|||\cdot|||$ norm, interpolating U on G_h. Clearly, v_h is a solution of (10.44). Now, by (10.52) for $j = 0$,

$$\|U-v_h\|_{L^2} \leq \|U-SU\|_{L^2} + \|Sv_h-v_h\|_{L^2}$$
$$\leq C'h^m[\|U\|_m+\|v_h\|_m].$$

Now utilize (10.46i) with $u = v_h$ and recall that v_h interpolates U and that $B(v_h,v_h) \leq B(U,U)$. Thus,

$$(10.53)\quad \|U-v_h\|_{L^2} \leq C'h^m[\|U\|_m+\{Ch^n \sum_{x \in G_h} [U(x)]^2+B(U,U)\}^{1/2}]$$

and (10.51) follows upon applying (10.46i), with u = U, to (10.53). Thus we have established (10.51) for j = 0. The result for $0 < j < m$ follows from standard interpolation theorems (cf. [1.1, Lemma 13.3]) with no assumption on the geometry of Ω. □

10.3 Approximation by Extended L^1 Extremals

In this section, we examine the approximation properties of the solutions of the extended extremal problem considered in §5. Our major result is the following.

Theorem 10.14. Let $f \in W^{n,1}(I)$, I a compact interval in $\mathbb{R}$, with $\|D^n f\|_{L^1(I)} \leq 1$ and $D^j f(a) = 0$ for $0 \leq j \leq n-1$. Let a partition $\Delta : a = x_0 < \dots < x_m = b$ of I be specified with $m \geq n-1$ such that Δ contains a fixed set $\{Y_1,\dots,Y_n\}$ of n points. Let s be the spline function, guaranteed by Corollary 5.3, agreeing with f on Δ. Then there exists a positive constant C, independent of f and Δ, such that

$$(10.54)\quad \|D^j(f-s)\|_{L^r(I)} \leq C\bar{\Delta}^{n-1-j},\ 0 \leq j \leq n-1,\ \text{for } 1 \leq r \leq \infty.$$

Moreover, the order given by (10.54) is sharp for n = 2, j = 0,

$r = 1$ and $I = [0,1]$: for each sufficiently small $\varepsilon > 0$,

$$(10.55) \qquad \|\bar{\varepsilon}x^{1+\varepsilon} - s_m\|_{L^1(I)} \geq C_\varepsilon\left(\frac{1}{m}\right)^{1+\varepsilon}, \quad C_\varepsilon > 0,$$

where $s_m(i/m) = \bar{\varepsilon}(i/m)^{1+\varepsilon}$, $i = 0,\ldots,m$; here s_m is the spline interpolant of Corollary 5.3, $m = 1,2,\ldots$, and $\bar{\varepsilon} = (1+\varepsilon)^{-1}$.

<u>Lemma 10.15</u>. Let f, Δ and s be given as in Theorem 10.14. There exists a positive constant C', independent of f and Δ, such that

$$\|D^{n-1}(f-s)\|_{L^\infty} \leq C'.$$

<u>Proof</u>. Since

$$f^{(n-1)}(x) = \int_a^x f^{(n)}(t)\,dt$$

it follows that $\|D^{n-1}f\|_{L^\infty(I)} \leq 1$. Since

$$\|D^{n-1}s\| \leq |D^{n-1}s(a+)| + 1,$$

it suffices to show that $|D^{n-1}s(a+)|$ is bounded independently of f and Δ. Now s may be written

$$(10.56) \qquad s(x) = P(x) + \sum_{j=0}^{r} a_j(x-t_j)_+^{n-1}$$

where $\sum_{j=0}^{r} |a_j| \leq 1$, P is a polynomial of degree $n - 1$ and $a < t_0 < \ldots < t_r < b$. Thus, $D^{n-1}s(a+) = D^{n-1}P(a)$. Now since the space $\mathscr{P}_{n-1}$ of real polynomials of degree $n - 1$ is of finite dimension n, it follows that any two norms are equivalent. In particular, any set of such polynomials bounded in the norm

$$(10.57)\qquad \|P\|_{\mathscr{P}_{n-1}} = \sum_{i=1}^{n} |P(Y_i)|$$

is bounded in the norm

$$(10.58)\qquad |||P|||_{\mathscr{P}_{n-1}} = \sum_{j=0}^{n-1} |c_j|,\ P(x) = \sum_{j=0}^{n-1} c_j x^j.$$

We shall consider the set of polynomials P, given by (10.56), as f and Δ vary according to the hypotheses of Theorem 10.14. Now on the point set $\{Y_1,\ldots,Y_n\}$, f and s agree so that

$$(10.59)\qquad P(Y) = f(Y) - \sum_{j=0}^{r} a_j (Y-t_j)_+^{n-1}$$

if $Y \in \{Y_1,\ldots,Y_n\}$. Now the representation

$$f(x) = \int_a^b \{(x-t)_+^{n-1}/(n-1)!\} f^{(n)}(t)\,dt$$

implies that $\|f\|_{C[a,b]} \leq (b-a)^{n-1}/(n-1)!$ and, clearly, the function

$$g(x) = \sum_{j=0}^{r} a_j (x-t_j)_+^{n-1}$$

satisfies

$$\|g\|_{C[a,b]} \leq (b-a)^{n-1}$$

so that the set of polynomials P, given by (10.56), is bounded in the norms (10.57) and (10.58). In particular, the coefficients c_{n-1} are uniformly bounded so that the numbers

$$D^{n-1}s(a+) = D^{n-1}P(a) = (n-1)!c_{n-1}$$

are uniformly bounded. This completes the proof of the lemma. □

<u>Proof of Theorem 10.14</u>. Standard arguments employing Rolle's Theorem (cf. eq. (10.6)) give the inequalities

$$(10.60)\quad \|D^j(f-s)\|_{L^r(I)} \leq \overline{\Delta}(j+1)\|D^{j+1}(f-s)\|_{L^r(I)},\quad 0 \leq j \leq n-2.$$

Combining (10.60) with Lemma 10.15, we obtain (10.54) with $C = C'(j+1)\ \ldots\ (n-1)(b-a)^{1/r}$.

To obtain (10.55), we first observe that the knots $t_0,\ldots,t_r$ of s_m occur precisely at subset of the interior nodal points $x_1,\ldots,x_{m-1}$. To show this, suppose first that no three adjacent graph points $(x_i,f(x_i))$ are collinear. Then, setting $J_i = [x_i,x_{i+1}]$, $i = 0,\ldots,m-1$, we see that, either $t_i \in J_i$, or $t_i \in J_{i+1}$. In either case, the magnitude of the difference in slopes of two linear functions, interpolating at the nodes contained in $J_i \cup J_{i+1}$ such that their graphs form a sawtooth on $J_i \cup J_{i+1}$ and meet at a point distinct from $(x_i,f(x_i))$, must strictly exceed the magnitude of the difference in slopes for the sawtooth function on $J_i \cup J_{i+1}$ with knot at x_i. The case of collinear graph points is similar. Having established that the knots of s_m are a subset of the nodes, (10.55) is a simple consequence of an adaptation of an argument first used by Schultz and Varga [10.6], as we now show.

Now fix $0 < \epsilon < 1$ and let

$$0 < \tau(\epsilon) = \inf_{P \in \mathscr{P}_1} \{\|\overline{\epsilon}x^{1+\epsilon}-P\|_{L^1(0,1)}\}.$$

Then, for the interval $J_0 = [0,h]$, we have, after a change of variable $x = hy$,

$$\|\bar{\epsilon}x^{1+\epsilon}-s_m\|_{L^1(J_0)} \geq h^{2+\epsilon}\tau(\epsilon).$$

Precisely the same analysis yields

$$\|\bar{\epsilon}x^{1+\epsilon}-s_m\|_{L^1(J_i)} \geq h^{2+\epsilon}\tau(\epsilon),\ i = 1,\ldots,m-1$$

for $J_i = [ih,(i+1)h]$. Thus,

$$\begin{aligned}\|\bar{\epsilon}x^{1+\epsilon}-s_m\|_{L^1(0,1)} &= \sum_{i=0}^{m-1} \|\bar{\epsilon}x^{1+\epsilon}-s_m\|_{L^1(J_i)} \\ &\geq mh^{2+\epsilon}\tau(\epsilon) \\ &= h^{1+\epsilon}\tau(\epsilon).\end{aligned}$$

Now the function $f(x) = \bar{\epsilon}x^{1+\epsilon}$ satisfies $f \in W^{2,1}(0,1)$ with $\|f''\|_{L^1(0,1)} = 1$ and $f(0) = f'(0) = 0$. Thus, choosing $Y_0 = 0$ and $Y_1 = 1$ we see that f and Δ satisfy the hypotheses of the first part of Theorem 10.14. This concludes the proof of the theorem. □

Remarks. The reader will observe that the order of approximation in (10.54), which is seen to be best possible in a generic sense for this approximation process, is of order one less than that achievable by optimal linear approximation processes (cf. the linear approximation process, defined for splines of degree $n - 1$, constructed in [10.4] and valid for $W^{n,1}$ functions). The explanation for this appears to be that the approximation process defined by Theorem 10.14 is actually intended for the larger class of functions whose nth. derivatives, in the sense of measures, have variation (as measures) not exceeding one.

The results of section 10.1 are taken from Jerome [1.6]. The existence Theorem 10.1, for $p = \infty$, represents an improvement over corresponding results in [1.6] where certain hypotheses, now seen to be unnecessary, were made. The idea of using results on k-widths to provide lower bounds for approximation by interpolating extremals first appeared in [1.6]. This idea was carried further by Jerome and Schumaker [10.4].

The convergence results of section 10.1 are of the same order for nonlinear T as for linear T. The convergence results of section 10.2 roughly assert that any function U in $W_0^{m,2}(\Omega)$, Ω a bounded open set in $\mathbb{R}^n$, can be approximated to order h^m by a correspondingly smooth function v_h, interpolating U on the points of a uniform grid of side h, in such a way that v_h solves an elliptic boundary value problem in $\Omega - G_h$. The reader may observe that the general hypotheses of section 10.2 are satisfied for most bilinear forms in one Euclidean variable (cf. [3.4]), even those which are nonsymmetric.

The results of section 10.2 are excerpted from the authors' paper [4.1] while the results of section 10.3 are contained in [5.1].

REFERENCES

10.1 J. P. Aubin, Approximation of Elliptic Boundary Value Problems, Wiley-Interscience, New York, 1972.

10.2 J. H. Bramble and S. R. Hilbert, "Estimation of linear functionals on Sobolev spaces with application to Fourier transforms and spline interpolation," SIAM J. Numer. Anal., 7 (1970), 112-124.

10.3 A. El Kolli, "Nieme epaisseur dans les espaces de Sobolev," J. Approximation Theory, 10 (1974), 268-294.

10.4 J. W. Jerome and L. L. Schumaker, "On the distance to a class of generalized splines," ISNM 25, Birkhäuser Verlag, Basel, 1974, 503-517

10.5 G. G. Lorentz, Approximation of Functions, Holt, Rinehart and Winston, New York, 1966.

10.6 M. H. Schultz and R. S. Varga, "L-splines," Numer. Math., 10 (1967), 345-369.

PART VII. PERFECT SPLINE SOLUTIONS IN THE THEORY OF BEST APPROXIMATION IN L^∞

§11. The Trigonometric and Algebraic Favard Problem

11.1 Function Class Approximation by Trigonometric Polynomials

This section contains the well-known theorem on best approximation by trigonometric polynomials, a theorem discovered simultaneously by J. Favard [11.4] and N. I. Achieser and M. G. Krein [11.1] which is a sharpening of D. Jackson's theorem.

We denote by W_n^* the set of all those functions f on $[-\pi,\pi]$ for which $f,f',\dots,f^{(n-1)}$ are continuous and 2π-periodic and $|f^{(n-1)}(x)-f^{(n-1)}(y)| \leq |x-y|$ for all $x, y \in [-\pi,\pi]$. T_m is the space of (real) trigonometric polynomials of degree m; that is, T_m is the linear span of $\{\cos kx\}_{k=0}^m$ and $\{\sin kx\}_{k=1}^m$.

<u>Theorem 11.1</u>. Let $m \geq n - 1$. Then

$$(11.1) \qquad \beta_{nm} = \sup_{f \in W_n^*} \inf_{s \in T_m} \|f-s\|_{L^\infty(-\pi,\pi)} = K_n(m+1)^{-n}$$

where

$$(11.2) \qquad K_n = \frac{4}{\pi} \sum_{j=0}^{\infty} (-1)^{j(n+1)} (2j+1)^{-n-1}.$$

Furthermore, there is a solution f_0 of (11.1) such that $f_0^{(n)}(x) = \operatorname{sign}(\cos(m+1)x)$ if n is even, $f_0^{(n)}(x) = \operatorname{sign}(\sin(m+1)x)$ if n is odd. If g_0 is any other solution to (11.1), then $g_0(x) = \lambda f_0(x-x_0) + c$ where $\lambda = \pm 1$, $x_0 \in [-\pi,\pi]$, and c is a constant.

<u>Proof</u>. Although this theorem is well-known and readily available (see, for example [11.7, p. 115] we present a proof, both to be complete and to give some techniques which will be used later in

this section. We introduce the notation $E_m^*(g)$ as the distance in $L^\infty(-\pi,\pi)$ of the continuous periodic function g to the linear space T_m. We shall make use of the fact that the dual space of the continuous periodic functions on $[-\pi,\pi]$ is the space of finite regular Borel measures μ on $[-\pi,\pi]$ which satisfy $\mu(-\pi) = \mu(\pi)$.

We do the proof for n even; the case for n odd requires only minor modifications.

Let $D_n(x) = \frac{1}{2\pi}(-1)^{n/2} \sum_{\substack{k=-\infty \\ k\neq 0}} |k|^{-n} e^{ikx}$. Then for each $f \in W_n^*$ with mean value zero we have

$$f(t) = \int_{-\pi}^{\pi} f^{(n)}(x) D_n(t-x)\,dx.$$

Note that $\int_{-\pi}^{\pi} f^{(n)}(x)\,dx = 0$ since $f^{(n-1)}(\pi) = f^{(n-1)}(-\pi)$. Furthermore, if h is any function in the unit ball of L^∞ with mean-value 0, then there is a unique function $H \in W_n^*$ with mean-value zero and $H^{(n)} = h$; H is just the convolution of h and D_n.

Let λ be any (real) periodic measure on $[-\pi,\pi]$ of total variation at most one which is orthogonal to T_m and let $f \in W_n^*$ with mean value zero. Then

$$\beta_{nm} \geq \left|\int_{-\pi}^{\pi} f(t)\,d\lambda(t)\right|$$

$$= \left|\int_{-\pi}^{\pi} f^{(n)}(x) D_\lambda(x)\,dx\right|$$

where

$$D_\lambda(x) = \int_{-\pi}^{\pi} D_n(t-x)\,d\lambda(t).$$

Now let λ vary over all measures orthogonal to T_m of total variation at most one and let $f^{(n)}$ vary over all functions in the unit ball of L^∞ with mean-value zero. We find that

$$\beta_{nm} \geq \sup_{\lambda}\{\text{distance in } L^1 \text{ from } D_\lambda \text{ to the constants}\}.$$

Actually, this supremum is a maximum and equality holds since both W_n^* and the unit ball of the space of measures are compact. Now we choose a specific λ; λ consists of $2m + 3$ point masses at the points $-\pi + k\pi/m+1$, $k = 0,1,\ldots,2m + 2$ with weights $-1/(4m+4)$, $+1/(2m+2)$, $-1/(2m+2),\ldots,+1/(2m+2)$, $-1/(4m+4)$, respectively. For this λ, a simple computation shows

$$D_\lambda(x) = (m+1)^{-n}D_n((m+1)x)$$

since λ is orthogonal to $\cos kx$ unless k is a multiple of $m + 1$ in which case the integral has the value 1. Hence,

$$\beta_{nm} \geq (m+1)^{-n}[\text{distance of } D_n((m+1)x) \text{ to } \mathbb{R} \text{ in } L^1].$$

Note, however, that the L^1 distance of $D_n((m+1)x)$ to the constants is the same as the distance of $D_n(x)$ to the constants by periodicity and that this number is 4 times

$$\int_0^{\pi/2} D_n(x)\,dx = \frac{1}{\pi}\sum_{k=0}^{\infty} (2k+1)^{-n-1}(-1)^k.$$

Hence,

$$(m+1)^n\beta_{nm} \geq \frac{4}{\pi}\sum_{k=0}^{\infty} (-1)^k(2k+1)^{-n-1} = K_n.$$

This proves the inequality in one direction. The other direction is somewhat easier but requires some special properties of the kernel D_n.

Let S be the best L^1 approximation to D_n from T_m; then because D_n has property A_m (cf. [11.7]) we know that

$$\|D_n - S\|_1 = (m+1)^{-n} K_n;$$

see [11.7; p. 114] for the details. If $f \in W_n^*$, then

$$f(x) - s(x) = \int_{-\pi}^{\pi} f^{(n)}(t)[D_n(x-t) - S(x-t)]dt$$

where $s \in T_m$. Thus, $\beta_{nm} \leq \|D_n - S\|_1$ so that we must have

$$\beta_{nm} = (m+1)^{-n} K_n.$$

Suppose now that $F \in W_n^*$ and that the distance of F from T_m is β_{nm}. Then

$$F(x) - s(x) = \int_{-\pi}^{\pi} F^{(n)}(t)[D_n(x-t) - S(x-t)]dt$$

where $s \in T_m$ so that

$$\begin{aligned} \beta_{nm} &\leq \|F - s\|_\infty = |F(x_0) - s(x_0)| \\ &\leq \|F^{(n)}\|_\infty \|D_n(x_0 - t) - S(x_0 - t)\|_1 \\ &\leq \beta_{nm}. \end{aligned}$$

Hence,

$$F^{(n)}(D_n(x_0-t)-S(x_0-t)) \geq 0$$

a.e. and $|F^{(n)}| = 1$ where $D_n - S \neq 0$. A translation allows us to assume that $x_0 = 0$. However, $D_n - S$ changes sign at the points $k\pi/(m+1)$, $k = 0, \pm 1, \ldots, \pm m$ and only there, so that F must have the indicated form. □

<u>Corollary 11.2</u>. Let r be a positive integer and let U_r consist of all functions f in W_n^* for which $\hat{f}(k) = 0$ when $|k| \leq r$; then

$$\max_{f \in U_r} \|f\|_\infty = (r+1)^{-n} K_n .$$

<u>Proof</u>. Let H be a function in U_r which attains the maximum value of the left hand side. If Q is any element in T_r, then the convolution of $H^{(n)}$ and Q is identically zero. Hence,

$$H(x) = \int_{-\pi}^{\pi} H^{(n)}(t)[D_n(x-t)-Q(x-t)]dt$$

so that

$$\|H\|_\infty \leq \text{distance of } D_n \text{ to } T_r \text{ in } L^1$$

$$= (r+1)^{-n} K_n$$

by Theorem 11.1. On the other hand, the function

$$G(x) = (r+1)^{-n} \frac{4}{\pi} \sum_{k=0}^{\infty} (-1)^{k(n+1)} (2k+1)^{-n-1} \cos[(2k+1)(r+1)x]$$

is in U_r and satisfies $\|G\|_\infty = |G(0)| = (r+1)^{-n} K_n$. □

11.2. Approximation by Algebraic Polynomials.

Let I be the closed interval $[-1,1]$ and let π_m denote the space of algebraic polynomials of degree m or less. Let W_n consist of all functions f on I for which $f, f', \ldots, f^{(n-1)}$ are absolutely continuous and $|f^{(n)}| \leq 1$ a.e. Let

$$(11.3) \qquad \alpha_{nm} = \sup_{f \in W_n} E_m(f)$$

where $E_m(f)$ is the distance of the continuous function f to the linear space π_m. In this section we prove two theorems. The first describes a property of any solution of (11.3); the second is a simple proof of a theorem of S. N. Bernstein on the asymptotic behavior of α_{nm} as $m \to \infty$.

<u>Theorem 11.3</u>. Let f be a solution of (11.3) where $m \geq n - 1$. Then $f^{(n)}$ assumes only the values 1 and -1 and has exactly $m - n + 1$ sign changes in $(-1,1)$. If $m = n - 1$, then f is a constant multiple of the nth Chebyshev polynomial. In all cases, if f is a solution of (11.3) then $f + P$ is a solution for any polynomial P of degree $n - 1$.

<u>Proof</u>. The proof closely resembles the proof of the first part of Theorem 11.1 and we use the fact that the dual space of the continuous functions on I is the space of finite regular Borel measures on I.

If $f \in W_n$ and if $f^{(\nu)}(-1) = 0$ for $\nu = 0, \ldots, n - 1$, then

$$f(x) = \int_{-1}^{1} f^{(n)}(t)\theta(x,t)\,dt$$

where $\theta(x,t) = (x-t)_+^{n-1}/(n-1)!$; that is, $\theta(x,t)$ equals $(x-t)^{n-1}/(n-1)!$ for $-1 \leq t \leq x$ and 0 for $x \leq t \leq 1$. Further, if h is in the unit ball of $L^\infty(I)$, then $H(x) = \int_I h(t)\theta(x,t)\,dt$ is in W_n.

Let λ be a (real) measure on I of total variation at most one which is orthogonal to π_m; let $f \in W_n$. We may assume $f^{(\nu)}(-1) = 0$ for $\nu = 0,\dots,n-1$. Since $m \geq n-1$ and λ is orthogonal to π_m we have

$$\begin{aligned}\alpha_{nm} &\geq \left|\int_I f(x)\,d\lambda(x)\right| \\ &= \left|\int_I f^{(n)}(t)F_\lambda(t)\,dt\right|\end{aligned}$$

where $F_\lambda^{(\nu)}(-1) = 0$, $0 \leq \nu \leq n-1$ and $F_\lambda^{(n)} = \lambda$ in the sense of distributions; that is,

$$F_\lambda(t) = \int_I \theta(x,t)\,d\lambda(x).$$

As λ runs over the measures of total variation at most one which are orthogonal to π_m and as f runs over W_n we find that

$$\alpha_{nm} \geq \sup_\lambda \|F_\lambda\|_{L^1(I)}. \tag{11.4}$$

The supremum is actually a maximum and equality holds in (11.4) for a measure λ with $m+2$ points in its support. To see this let H be a function in W_n whose distance to π_m is α_{nm} and for which $H^{(\nu)}(-1) = 0$, $0 \leq \nu \leq n-1$. Such a function exists since W_n is compact. Let $P \in \pi_m$ be the polynomial of degree m which is closest to H. Then there must be points $-1 \leq x_0 < x_1 \dots < x_{m+1} \leq 1$ at which $H(x_k) - P(x_k) = \alpha_{nm}(-1)^k$, $k = 0,\dots,m+1$. Let λ be a measure supported on $\{x_k : k = 0,\dots,m+1\}$ of total mass one which is orthogonal to π_m; let λ_k be the weight of λ at x_k, $k = 0,\dots,m+1$; it is easy to see that $(-1)^k\lambda_k > 0$ and hence

$\sum_{0}^{m+1} (-1)^k \lambda_k = 1$. An explicit calculation of the λ_k reveals that they are the normalized (m+1)st. order divided difference coefficients. We have now

$$\alpha_{nm} = \sum_{k=0}^{m+1} (H(x_k) - P(x_k))\lambda_k$$

$$= \int_I (H-P)\,d\lambda = \int_I H d\lambda$$

$$= \int_I H^{(n)} F_\lambda dx$$

$$\leq \|F_\lambda\|_1 \leq \alpha_{nm}.$$

Hence, equality holds in (11.4). We have also shown that if $H \in W_n$ is at distance α_{nm} from π_m with $H^{(\nu)}(-1) = 0$, $0 \leq \nu \leq n - 1$ then there is a measure λ, depending on H, with m + 2 points in its support which is orthogonal to π_m and which satisfies

$$\alpha_{nm} = \int_I H^{(n)} F_\lambda .$$

Hence, $H^{(n)} F_\lambda \geq 0$ a.e. and $|H^{(n)}| = 1$ where $F_\lambda \neq 0$. However, either F_λ or $-F_\lambda$ is the (m-n+1)st. derivative of the function

$$B(x) = \int_I (x-t)_+^m/m!\ d\lambda(t)$$

which is a B-spline and so F_λ has precisely m - n + 1 zeros in (-1,1) if $n \geq 2$; see [11.3; p. 74]. If n = 1, F_λ is a step-function with discontinuities at the $\{x_k\}$. In particular, in the case m = n - 1, the smallest value of m for which α_{nm} is finite, we find that $|F_\lambda| > 0$ on (-1,1) and so H is a polynomial of degree n;

clearly, H has the form $x^n + p(x)$ where p has degree n - 1 and hence

$$\alpha_{n,n-1} = 2^{-n+1}/n!. \qquad \square$$

For emphasis we restate the primary conclusion of Theorem 11.3: <u>Each solution of</u> (11.3) <u>is a perfect spline with exactly</u> m - n + 1 <u>knots on</u> [-1,1].

We now use Corollary 11.2 and several other facts to give a proof of a theorem of S. N. Bernstein [11.2], proved in 1947; an English language proof is in [11.9, p. 293].

<u>Theorem 11.4</u>. $\lim_{m \to \infty} m^n \alpha_{nm} = K_n$ where K_n is the constant given by (11.2).

<u>Proof</u>. For this proof it is convenient to deal with (a modification of) the kernel D_n rather than the kernel $\theta(x,t)$ of §11.2. Let

$$d_n(x) = D_n(\pi x)/\pi^{n-1}, \quad -1 \leq x \leq 1.$$

If h is in the unit ball of $L^\infty(I)$, then $H(x) = \int_I d_n(x-t)h(t)\,dt$ is in W_n and differs from $\int_I \theta(x,t)h(t)\,dt$ by a polynomial of degree n - 1. Hence, as in the proof of Theorem 11.3

$$\alpha_{nm} = \sup_\lambda \|d_\lambda\|_{L^1(I)}$$

where the supremum is taken over all measures λ on I which are orthogonal to π_m and which have total variation at most one.

First we establish that

$$\limsup_{m \to \infty} m^n \alpha_{nm} \leq K_n .$$

Let ϵ be a small positive number and let r be the greatest integer in $m/\pi(1+\epsilon)$; of course, r varies with m but we suppress this in the notation. We shall need to use the following standard fact [see 11.7; p. 77].

<u>Lemma 11.5</u>. Let $R > 1$ and let E_R be the ellipse $x = 1/2(R+R^{-1})\cos\theta$, $y = 1/2(R-R^{-1})\sin\theta$, $0 \leq \theta \leq 2\pi$. Suppose f is holomorphic on and within E_R and bounded by M on E_R. Then

$$E_m(f;I) \leq 2MR^{-m}(R-1)^{-1}. \tag{11.5}$$

Take $f(z) = e^{ikz\pi}$. The maximum of $|f(z)|$ on E_R is less than $\exp[k\, 1/2(R-R^{-1})\pi]$. When $0 \leq k \leq r$, this maximum is smaller than $\exp[m(R-R^{-1})/2(1+\epsilon)]$. Fix $R > 1$ sufficiently close to 1 so that

$$\exp[(R-R^{-1})/2(1+\epsilon)] = \rho R$$

where $\rho < 1$. Then we have the estimate

$$E_m(e^{ik\pi x};I) \leq 2(R-1)^{-1}\rho^m \tag{11.6}$$

for $0 \leq k \leq r$. Hence

$$E_m\Big(\sum_{\substack{-r \\ k\neq 0}}^{r} |k|^{-r} e^{ik\pi x}\Big) \leq 4(R-1)^{-1} m\rho^m. \tag{11.7}$$

Now let h be a function in the unit ball of L^∞. Let $\lambda \perp \pi_m$ and set

$$H_r(x) = \pi^{-n} \sum_{|k| > r} |k|^{-n}\hat{h}(k)\hat{\lambda}(k)\exp(ik\pi x)$$

where

$$\hat{h}(k) = 1/2\pi \int_I e^{-ik\pi t} h(t)\,dt$$

$$\hat{\lambda}(k) = 1/2\pi \int_I e^{-ik\pi t} d\lambda(t).$$

The nth. derivative of H_r differs from

$$\sum_{-\infty}^{\infty} \hat{h}(k)\hat{\lambda}(k)\exp(ik\pi x) \tag{11.8}$$

by the term

$$\sum_{|k| \le r} \hat{h}(k)\hat{\lambda}(k)\exp(ik\pi x). \tag{11.9}$$

However, (11.8) is just

$$\int_I h(x-t)\,d\lambda(t)$$

and this is no larger than 1 as a simple exercise shows. (We extend h to be periodic on $\mathbb{R}$.) Further, the estimate (11.6) implies that (11.9) is no larger than $4(R-1)^{-1}m\rho^m$. Thus, Corollary 11.2 implies, after a change of variables,

$$|H_r(0)| \le \|H_r\|_\infty \le \pi^{-n}(r+1)^{-n}K_n(1+4(R-1)^{-1}m\rho^m).$$

Thus,

$$\alpha_{nm} \le 4(R-1)^{-1}m\rho^m + (r+1)^{-n}\pi^{-n}K_n(1+4(R-1)^{-1}m\rho^m)$$

so that

$$\limsup_{m \to \infty} (m^n \alpha_{nm}) \leq K_n(1+\varepsilon)^n$$

Since ε is arbitrary, we have established

$$\limsup_{m \to \infty} m^n \alpha_{nm} \leq K_n.$$

To prove that $\liminf(m^n \alpha_{nm}) \geq K_n$, we shall use a few elementary facts about entire functions of exponential type. Let $\varepsilon > 0$ be given. Let

$$F_m(x) = (4/\pi) \sum_{0}^{\infty} (-1)^k (2k+1)^{-n-1} \cos(2k+1)(1+\varepsilon)mx, \quad -\infty < x < \infty$$

and let $F(x) = F_1(x)$. Suppose for each m in a sequence of $m \to \infty$ there is a polynomial p_m of degree m with

$$\|F_m - p_m\|_{L^\infty(I)} \leq (1-\delta)K_n, \quad \delta > 0,$$

where δ is independent of m. Then,

$$\|F(x) - q_m(x)\|_{L^\infty(-m,m)} \leq (1-\delta)K_n$$

where $q_m(x) = p_m(x/m)$, $-\infty < x < \infty$. Now

$$q_m^{(k)}(0) = m^{-k} p_m^{(k)}(0), \quad k = 0, \ldots, m$$

and by a classical inequality of V. A. Markov [11.8; p. 53]

$$|p_m^{(k)}(0)| \leq m^k \max_{|x| \leq 1} |p_m(x)|$$

$$\leq 2K_n m^k.$$

Hence,

$$|q_m^{(k)}(0)| \leq 2K_n \text{ for } k = 0,\ldots,m.$$

This implies that a subsequence of $\{q_m\}$ converges uniformly on compact subsets of the plane to an entire function G of exponential type 1 or less. Clearly, G satisfies the inequality

$$\|F-G\|_{(-\infty,\infty)} \leq (1-\delta)K_n$$

and we may assume that G is a best approximation of F. However, since F is periodic with period $2\pi(1+\varepsilon)^{-1}$ we may assume G has this period and hence G is constant since its type is one or less. But no constant is within distance K_n of F. This contradiction shows that

$$\liminf_{m \to \infty} E_m(F_m;I) \geq K_n.$$

However, $(1+\varepsilon)^{-n}m^{-n}F_m \in W_n$ so that

$$\liminf_{m \to \infty} (m^n \alpha_{nm}) \geq K_n(1+\varepsilon)^{-n}.$$

Thus,

$$\liminf_{m \to \infty} m^n \alpha_{nm} \geq K_n$$

and the theorem is proved. $\square$

11.3. Approximation on the Line by Entire Functions of Exponential Type

Let E_σ, $\sigma > 0$, be the space of entire functions of exponential type <u>less than</u> σ which are bounded on the real axis. Such a function f necessarily satisfies the growth condition

$$|f(x+iy)| \leq e^{\rho|y|}\left(\sup_{-\infty < t < \infty} |f(t)|\right)$$

for some $\rho < \sigma$. In this section we use Theorems 11.1 and 11.4 to give a simple proof of the following theorem of M. G. Krein [11.6].

<u>Theorem 11.6</u>. Let V_n consist of all bounded function f on $(-\infty,\infty)$ which satisfy $|f^{(n)}| \leq 1$ on $(-\infty,\infty)$. Let

$$\gamma_{n\sigma} = \sup_{f \in V_n} \inf_{G \in E_\sigma} \|f-G\|_{(-\infty,\infty)}.$$

Then $\gamma_{n\sigma} = \sigma^{-n}K_n$ where K_n is the constant given in (11.2).

<u>Proof</u>. Again, we take n to be even. Let

$$F_\sigma(x) = \frac{1}{\sigma^n}\frac{4}{\pi}\sum_{k=0}^{\infty} (-1)^k \frac{\cos(2k+1)\sigma x}{(2k+1)^{n+1}}, \quad -\infty < x < \infty.$$

Then $F_\sigma \in V_n$; suppose $G \in E_\sigma$ and $\|G-F_\sigma\| \leq K_n(1-\delta)\sigma^{-n}$, $\delta > 0$. Let $F(x) = \sigma^n F_\sigma(x/\sigma)$ and $H(z) = \sigma^n G(z/\sigma)$. Then $\|F-H\|_{(-\infty,\infty)} \leq (1-\delta)K_n$ and H is entire of exponential type less than 1. Since F is 2π-periodic, we may assume H is also and hence H is constant. But the distance from F to the constants is K_n. Thus, the distance from F_σ to E_σ is $K_n\sigma^{-n}$ so that $\gamma_{n\sigma} \geq K_n\sigma^{-n}$.

On the other hand, let $f \in V_n$, let $\epsilon > 0$, and let I_m be the interval $I_m = [-m/\sigma(1-\varepsilon), m/\sigma(1-\varepsilon)]$. Let p_m be the best approximation to f on I_m from π_m, let $g_m(x) = f(mx/\sigma(1-\varepsilon))$ and $q_m(x) = p_m(mx/\sigma(1-\varepsilon))$. Then

$$\begin{aligned} \|f-p_m\|_{I_m} &= \|g_m-q_m\|_I \\ &= E_m(g_m;I) \\ &\leq \alpha_{nm} m^n/\sigma^n(1-\varepsilon)^n \end{aligned}$$

so that by Theorem 11.4

$$\limsup_{m\to\infty} E_m(f;I_m) \leq K_n/\sigma^n(1-\varepsilon)^n.$$

Hence, $\|q_m\|_I \leq C$ for all m so that

$$\begin{aligned} |p_m^{(k)}(0)| &= \sigma^k(1-\varepsilon)^k m^{-k}|q_m^{(k)}(0)| \\ &\leq C\sigma^k(1-\varepsilon)^k \end{aligned}$$

by V. A. Markov's theorem. Hence, some subsequence of $\{p_m\}$ converges uniformly on compact subsets of the plane to an entire function G of exponential type less than σ which must satisfy

$$\|f-G\|_{(-\infty,\infty)} \leq K_n/\sigma^n(1-\varepsilon)^n.$$

Hence,

$$\inf_{G\in E_\sigma} \|f-G\|_{(-\infty,\infty)} \leq K_n/\sigma^n(1-\varepsilon)^n$$

for each $f \in V_n$ and each $\varepsilon > 0$, so that

$$\gamma_{n\sigma} \leq K_n/\sigma^n$$

and this completes the proof. □

To complete the circle of ideas in Theorems 11.1, 11.4, and 11.6 we show that Theorem 11.6 easily implies the value of the constant β_{nm} in Theorem 11.1. Let $f \in W_n^*$ and extend f periodically to the line. Then, because a 2π-periodic function in E_{m+1} is a trigometric polynomial of degree m or less, we have

$$\inf_{G \in E_{m+1}} \|f-G\|_{(-\infty,\infty)} = \inf_{T \in T_m} \|f-T\|_{L^\infty(-\pi,\pi)} .$$

But the left-handed side does not exceed $K_n(m+1)^{-n}$ by Theorem 11.6. Hence, $\beta_{nm} \leq K_n(m+1)^{-n}$. On the other hand, the function

$$F_m(x) = (m+1)^{-n} \frac{4}{\pi} \sum_0^\infty (-1)^{k(n+1)} (2k+1)^{-n-1} \cos[(2k+1)(m+1)x]$$

lies in W_n^* and has 2m + 3 alternation on $[-\pi,\pi]$ and hence the best approximation to F_m from T_m is zero; thus

$$\beta_{nm} \geq \|F_m\|_\infty \geq |F_m(0)| = (m+1)^{-n} K_n .$$

<u>Remarks</u>. The present chapter, showing the essential equivalence among the Achieser-Favard-Krein theorem, on best approximation of the periodic class $\{f : |f^{(n)}| \leq 1\}$ by trigonometric polynomials of degree m, the asymptotic Bernstein theorem, on the behavior of the best constants in the corresponding algebraic problem as $m \to \infty$, and the Krein theorem, on best approximation by entire functions of prescribed exponential type, is excerpted from the paper of Fisher [11.5]. We emphasize that the extremals in the trigonometric and algebraic approximation problems as well as the Krein theorem on $\mathbb{R}$,

are perfect spline functions. The result for the algebraic problem is new and gives the precise number of knots for the extremals.

REFERENCES

11.1 N. I. Achieser and M. G. Krein, "On the best approximation of periodic functions," Doklady Akad. Nauk SSSR, 15 (1937), 107-112.

11.2 S. N. Bernstein, Collected Works, vol. II, Akad. Nauk SSSR, Moscow, 1954.

11.3 H. B. Curry and I. J. Schoenberg, "On Polya Frequence functions I," J. d'Analyse Math. XVII (1966), 71-108.

11.4 J. Favard, "Sur les meilleures procédés d'approximation de certaines classes des fonctions par des polynômes trigonométriques," Bull. Sci. Math., 61 (1937), 209-224, 243-256.

11.5 S. D. Fisher, "Best approximation by polynomials," submitted for publication.

11.6 M. G. Krein, "On the approximation of continuous differentiable functions on the whole real axis," Doklady Akad. Nauk., 18 (1938)

11.7 G. G. Lorentz, Approximation of Functions, Holt, Rinehart and Winston, New York, 1966.

11.8 I. P. Natanson, Constructive Theory of Functions, A. E. C. translation series 4503.

11.9 A. F. Timan, Theory of Approximation of Functions of a Real Variable, translated by J. Berry, Pergamon Press Ltd., Oxford, England, 1963.

§12. Minimization and Interpolation at Integer Points of the Real Axis

12.1 The Fundamental Interval of Uniqueness for Periodic Data

Let $\mathbb{Z}$ denote the integers and let $\underline{a} = (a_i)$ be a doubly infinite bounded set of reals. Let

$$(12.1) \qquad U = \{f \in W^{n,\infty}(\mathbb{R}) : f(i) = a_i,\ i \in \mathbb{Z}\}$$

and

$$(12.2) \qquad \alpha = \inf\{\|f^{(n)}\|_\infty : f \in U\}$$

We wish to investigate existence and uniqueness questions for solutions to (12.2). We can not appeal directly to the earlier theorems for characterization since the set $U_0 = \{g \in W^{n,\infty}(\mathbb{R}) : g(i) = 0,\ i \in \mathbb{Z}\}$ does not have finite codimension in $W^{n,\infty}(\mathbb{R})$. Nonetheless, we shall show that many of the same results hold when the sequence $\{a_i\}$ is periodic, an important special case needed in the sequel.

We begin with two lemmas needed in later constructions.

<u>Lemma 12.1</u>. Let A and B be two compact sets of positive measure with $a < b$ for every $a \in A$ and $b \in B$. Let $x_1 < \dots < x_n$ be n points strictly between A and B and let $r_1,\dots,r_n$ be real numbers. Then there is an $h \in W^{n,\infty}(\mathbb{R})$ such that

(i) $D^n h$ is supported on $A \cup B$

(12.3) (ii) $h(x_i) = r_i$ for $1 \le i \le n$

(iii) $h \equiv 0$ to the left of A and to the right of B

Proof. Let $a_0 = \inf\{a : a \in A\}$, $b_0 = \inf\{b : b \in B\}$ and $b_1 = \sup\{b : b \in B\}$. Choose a function h_1 such that $h_1(x_i) = r_i$, $1 \leq i \leq n$, $h_1 \equiv 0$ on $(-\infty, a_0]$ and $D^n h_1$ is supported on A. Such an h_1 exists by the argument of Lemma 6.5. Next choose an h_2 such that $D^n h_2$ is supported on B, $h_2 \equiv 0$ on $(-\infty, b_0]$, and

$$\int_B D^n h_2(t)(b_1 - t)^j dt = - \int_A D^n h_1(t)(b_1 - t)^j dt$$

for $j = 0, 1, \ldots, n - 1$. Again it is elementary that such an h_2 exists. Now let $h = h_1 + h_2$. Then h clearly satisfies (i) and (ii) above as well as the first part of (iii). On $[b, \infty)$ we know that $D^n h \equiv 0$; also for $0 \leq j \leq n - 1$

$$(n-1-j)!\,h^{(j)}(b_1) = \int_{a_0}^{b_1} D^n h(t)(b_1 - t)^{n-1-j} dt$$

$$= \int_{A \cup B} D^n h(t)(b_1 - t)^{n-1-j} dt$$

$$= 0$$

and hence $h \equiv 0$ on $[b_1, \infty)$. □

We shall say that two disjoint sets A and B in $\mathbb{R}$ intersperse at least k times if there are subsets $A_1, \ldots, A_{k_1}$ of A of positive measure, where $k_1 = 1 + [k/2]$, and subsets $B_1, \ldots, B_{k_1}$ of B if k is odd and $B_1, \ldots, B_{k_1 - 1}$ if k is even, of positive measure which satisfy the following inequalities for all j for which they are meaningful:

$$a_j < b_j < a_{j+1} \quad \text{for all } a_j \in A_j,\ b_j \in B_j$$

Lemma 12.2. For a given positive integer n, let $N = n + 1$ if n is odd and $N = n$ if n is even. Let A and B be disjoint compact sets in $(0,N)$ which intersperse $2n$ or more times if n is odd and $2n-1$ or more times if n is even, such that each of the open intervals $(k,k+n)$, $k = -n+1,\dots,N-1$, meet both A and B in a set of positive measure. Let $r_1,\dots,r_{N-1}$ and $d_0,\dots,d_{n-1}$ be arbitrary real numbers. Then there is a function $h \in C^{n-1}(\mathbb{R})$, with $h^{(n)} \in L^\infty(\mathbb{R})$, such that

(i) $h(j) = r_j$ for $j = 1,\dots,N-1$

(ii) $h \equiv 0$ on $(-\infty,0]$

(12.4) (iii) $D^n h$ is non-negative on A, non-positive on B and 0 off $A \cup B$

(iv) $h^{(j)}(N) = d_j$ for $j = 0,\dots,n-1$.

Proof. Let n be odd and let V consist of all $L^\infty(\mathbb{R})$ functions which vanish off $A \cup B$ and which are non-negative on A and non-positive on B. Let $M_1,\dots,M_{2n}$ denote those B-splines of degree $n-1$ with integer knots whose support intersects $(0,n+1)$; recall that such a B-spline of degree $n-1$ is the nth. divided difference of the function $n(x-\xi)_+^{n-1}$ taken with respect to the first argument on $n+1$ consecutive integers. We define a mapping $T : V \to \mathbb{R}^{2n}$ by

$$(12.5)\qquad Tg = \left(\left\{\int_{-\infty}^{\infty} g(t)\,\theta_j(t)\,dt\right\}_{j=1}^{n}, \left\{\int_{-\infty}^{\infty} g(t)\,\psi_j(t)\,dt\right\}_{j=0}^{n-1}\right)$$

Here, $\theta_j(t) = (j-t)_+^{n-1}/(n-1)!$ and $\psi_j(t) = (N-t)^{n-j-1}/(n-j-1)!$. Now, if $TV \neq \mathbb{R}^{2n}$, then the origin is not an interior point of the convex cone TV and there are scalars $\lambda_1,\dots,\lambda_n$, $\mu_0,\dots,\mu_{n-1}$, not all zero, such that

$$0 \leq \sum_{j=1}^{n} \lambda_j \int_{-\infty}^{\infty} g(t)\theta_j(t)\,dt + \sum_{j=0}^{n-1} \mu_j \int_{-\infty}^{\infty} g(t)\psi_j(t)\,dt$$

for all $g \in V$. Since each spline of degree $n - 1$ on $[0,n+1]$ can be represented as a linear combination of $M_1,\dots,M_{2n+1}$ it follows that there are scalars $a_1,\dots,a_{2n+1}$, not all zero, with

$$0 \leq \int_{-\infty}^{\infty} g(t)\Big[\sum_{j=1}^{2n+1} a_j M_j(t)\Big]dt$$

for all $g \in V$. Hence, if S is the function in the square brackets, then S is nonnegative on A and nonpositive on B. Since A and B intersperse at least $2n$ times and S is continuous for $n \geq 2$, S has at least $2n$ zeros, with at least one zero on each interval $(k,k+n)$, $-n+1 \leq k \leq N - 1$. It follows [12.4, p. 524, Lemma 4.2] that $a_1 = a_2 = \dots = a_{2n+1} = 0$ and hence $TV = \mathbb{R}^{2n}$. Thus, select g such that $Tg = (\{r_1\}_{j=1}^{n}, \{d_j\}_{j=0}^{n-1})$ and define

$h(x) = \int_{-\infty}^{\infty} \{(x-t)_+^{n-1}/(n-1)!\}g(t)\,dt$. Then h satisfies (12.4) and the proof is complete for n odd. If n is even the proof is identical except that T maps V into $\mathbb{R}^{2n-1}$ and only $2n-1$ B-splines are utilized. □

Theorem 12.3. (i) Let U be given by (12.1) and α by (12.2). Then there is a function $f \in U$ with $\|f^{(n)}\| = \alpha$. Furthermore, there is a solution f_0 of (12.2) with $|f_0^{(n)}| = \alpha$ a.e. on $\mathbb{R}$ and $f_0^{(n)}$ has at most n sign changes between successive integers.

(ii) In the case when the data $\{a_i\}$ is periodic with minimum period τ, then (12.2) has a solution with period τ. If $S(\tau)$ denotes the set of all solutions of (12.2) with period τ, then there is an interval of the form $[k,k+n]$ in $[0,\tau]$ (or $[0,\tau]$ itself if $\tau \leq n$) on which $|f^{(n)}| = \alpha$ a.e. for all $f \in S(\tau)$ and on which $f = g$ for all $f, g \in S(\tau)$. If $\tau \leq n$, then $S(\tau)$ contains precisely one element.

<u>Proof</u>. The existence of a solution to (12.2) follows routinely from Corollary 1.3 once it has been established that U is non-empty. This fact is a consequence, for example, of a cardinal spline interpolation theorem of Schoenberg [12.6]; bounded data may be interpolated uniquely by a bounded C^{n-1} spline of degree n; by the Markov-Bernstein theorem applied to each polynomial segment it follows that the (unique) cardinal spline interpolant has bounded nth. derivative and hence is in $W^{n,\infty}(\mathbb{R})$.

Let N be a positive integer and consider the set $U_N = \{f \in W^{n,\infty}(-N,N) : f(i) = a_i \text{ for } |i| \leq N\}$. Let

$$(*) \qquad \alpha_N = \inf\{\|f^{(n)}\|_\infty : f \in U_N\}$$

By 7.6 there is a solution f_N of (*) with $|f_N^{(n)}| \equiv \alpha_N$ and $f_N^{(n)}$ has n or fewer sign changes on any interval (i,i+1) in [-N,N]. Since any solution of (12.2) is an element of U_N and since $U_{N+1} \subset U_N$ for all N we know that $\alpha_1 \leq \alpha_2 \leq \dots \leq \alpha$; let $\lim \alpha_N = \beta$. We shall show that $\beta = \alpha$.

A diagonal argument produces a subsequence $\{f_{N_j}\}$ of $\{f_N\}$ such that $f_{N_j}^{(n)}$ converges in L^1 on each interval (i,i+1) to a function h with $|h| = \beta$ a.e. and h has n or fewer sign changes on (i,i+1). Let H be a function with $H^{(n)} = h$; after an appropriate polynomial of degree n - 1 is added to H, H lies in U and hence $\alpha \leq \beta$. Thus $\alpha = \beta$ and H is the desired solution.

To prove (ii) let f be any solution of (12.2) and put

$$f_N(x) = \frac{1}{N+1}\,(f(x) + f(x+\tau) + \dots + f(x+N\tau))$$

Then f_N is a solution of (12.2) and $|f_N(x)-f_N(x+\tau)| \leq C(N+1)^{-1}\|f\|_\infty$. The bounded sequence $\{f_N^{(n)}\}_{N=1}^\infty$ in $L^\infty(\mathbb{R})$ has a subsequence which is weak-$*$ convergent to a function $F^{(n)} \in L^\infty(\mathbb{R})$, where $F^{(k)}$, $0 \leq k \leq n-1$, is the uniform limit on compact subsets of $\mathbb{R}$ of corresponding derivatives of the subsequence. By arguments now familiar, F is a solution of (12.2) which is periodic with period τ. We shall show that $|F^{(n)}| = \alpha$ a.e. on some set of n consecutive intervals in $[0,\tau]$ (or if $\tau \leq n$, then $|F^{(n)}| = \alpha$ a.e. on $[0,\tau]$.)

Suppose there is a $\delta > 0$ such that the set $\Omega = \{|F^{(n)}| \leq \alpha - \delta\}$ meets any collection of n consecutive intervals in $[0,\tau]$ in a set of positive measure. A translation allows us to assume that $(0,1)$ contains a closed subset E_1 of Ω of positive measure. Let $E_2,\ldots,E_r$ be other closed subsets of Ω of positive measure lying in $(0,\tau)$, which do not contain any integer points and when arranged as (with obvious notation) $E_1 < E_2 < \ldots < E_r$, satisfy the condition that there are n or fewer integers between E_j and E_{j+1}, $j = 1,\ldots,r$, where E_{r+1} is E_1 translated to the right by τ. Such sets exist by the assumption that Ω meets any collection of n consecutive intervals in $[0,\tau]$ in a set of positive measure. (If $\tau < n$, then take $r = 1$.) By r applications of Lemma 12.1 there is a function $h \in W^{n,\infty}(\mathbb{R})$ with $h \equiv 0$ on $(-\infty,0] \cup [\tau+1,\infty)$, $h(i) = a_i$ for $i = 1,\ldots,\tau$, and $h^{(n)}$ is supported in $E_1 \cup \ldots \cup E_{r+1}$. Let $g(x) = \Sigma_{-\infty}^{\infty} h(x+k\tau)$. (Note that for each x at most 2 of these terms are non-zero.) Thus $g(i) = a_i$ for all $i \in \mathbb{R}$, $g^{(n)}$ is supported on the set Ω, and $\|g^{(n)}\|_\infty$ is finite. Hence, $(1+\varepsilon)^{-1}(F+\varepsilon g)$ is in U for all $\varepsilon > 0$ sufficiently small; this function satisfies $\|(1+\varepsilon)^{-1}(F^{(n)}+\varepsilon g^{(n)})\| \leq \alpha(1+\varepsilon)^{-1}$, a contradiction. Thus, $|F^{(n)}| = \alpha$ a.e. on some n consecutive intervals in $[0,\tau]$ or on

$[0,\tau]$ if $\tau \leq n$.

Now let $I_1,\ldots,I_p$ be those intervals of the form $[j,j+1]$ in $[0,\tau]$ for which there is a periodic solution $F_1,\ldots,F_p$, respectively, with $|F_j^{(n)}| < \alpha$ on a set of positive measure in I_j, $1 \leq j \leq p$. Then $F = p^{-1}(F_1 + \ldots + F_p)$ is a periodic solution and $|F^{(n)}| < \alpha$ on some subset of positive measure in each I_j, $1 \leq j \leq p$. Thus if $\tau \geq n$, there are n consecutive intervals in $[0,\tau]$ on which all periodic solutions G satisfy $|G^{(n)}| = \alpha$ a.e. We may assume $[0,n]$ is this set. Hence, if G, H are periodic solutions then $G^{(n)} = H^{(n)}$ on $[0,n]$ and since $G(i) = H(i)$ for $i = 0,\ldots,n$ we must have $G = H$ on $[0,n]$. Finally, note that if $\tau \leq n$, then $G = H$ on $\mathbb{R}$ so that $S(\tau)$ contains precisely one element. $\square$

12.2 The Euler Splines

In this section we examine the special case when the data satisfy the relation $a_i = (-1)^i$ and we show that (12.2) has, as its unique periodic solution, the Euler spline which has been explicitly constructed by Schoenberg [12.5]. Because of the earlier result of Schoenberg (cf. also Subbotin [12.8]) alluded to in section 12.1 on the unique interpolation of bounded data on $\mathbb{R}$ by bounded spline functions of degree n it would be sufficient to show that (12.2) possesses a unique solution which is a spline function of degree n in the case of $a_i = (-1)^i$; this bounded spline function must then be the Euler spline by the uniqueness of bounded spline interpolation. We shall not adopt this approach, however, and for the sake of completeness shall derive independently the properties of the Euler spline required in §13.

<u>Theorem 12.4</u>. Let the flat U in (12.1) be defined by the constants

$a_i = (-1)^i$, $i \in \mathbb{Z}$. Then there is a unique periodic solution E of (12.2) and E satisfies the following:

(i) E is a perfect spline function of degree n which is periodic of period 2;

(ii) $\|E\|_\infty = 1$

(iii) E has knots precisely at the integers if n is odd, and at the half-integers if n is even;

(iv) E is strictly monotone on each interval $[k,k+1]$, $k \in \mathbb{Z}$;

(v) E is even about each integer and odd about each half integer.

Proof. If $n \geq 2$, then Theorem 12.3 gives a unique periodic solution $E(x)$ with period 2.

If $n = 1$, then by Theorem 12.3 all periodic solutions agree say on $[0,1]$. Hence, $E(x) = -E(x+1)$ for $x \in [0,1]$ and if F is also a periodic solution, then $-F(x+1) = F(x) = E(x) = -E(x+1)$ for $x \in [0,1]$. Hence, $F = E$ on $[0,2]$ and hence $F \equiv E$ on $\mathbb{R}$ since both functions have period 2. Note that for $n = 1,2,\ldots$ $E(x) = -E(x+1)$ for the (unique) periodic solution E; in particular E is even about each integer.

We next show that $E^{(n)}$ changes sign at most twice on each interval of length 2. It suffices to consider $[0,2]$ since E has period 2. If $E^{(n)}$ has 3 or more sign changes on $[0,2]$, then since $E(x) = -E(x\pm 1)$, $E^{(n)}$ either has 2 sign changes on $(0,1)$, say at $0 < a < b < 1$, or one sign change in $(0,1)$, say at a, and another at 1. Note that if n is even $b \neq 1$ since $E^{(n)}$ is even about 1. Let

$$A = \{x \in \mathbb{R} : E^{(n)}(x) = -\alpha\}$$

$$B = \{x \in \mathbb{R} : E^{(n)}(x) = \alpha\}$$

and let N be as in Lemma 12.2. Then A and B intersperse at least 2n times on (0,n) if n is odd and 2n-1 times if n is even and each interval (k,k+n), k = -n+1,...,N - 1, contains subsets of A and B of positive measure. We shall use Lemma 12.2 to construct a function $F \in U$ with $\|F^{(n)}\| < \alpha$, a contradiction.

To construct F, let g satisfy

$$\text{(12.6)} \quad \begin{aligned} &\text{(i)} \quad g \equiv 0 \text{ on } (-\infty,0] \\ &\text{(ii)} \quad g^{(n)} = E^{(n)} \text{ on } [0,N] \\ &\text{(iii)} \quad g^{(n)} = 0 \text{ on } (N,\infty) \end{aligned}$$

According to Lemma 12.2 there is a function h with

$$\text{(12.7)} \quad \begin{aligned} &\text{(i)} \quad h \equiv 0 \text{ on } (-\infty,0] \\ &\text{(ii)} \quad h(i) = g(i) \text{ for } i = 1,\dots,N-1 \\ &\text{(iii)} \quad h^{(j)}(N) = g^{(j)}(N) \text{ for } j = 0,\dots,n-1 \\ &\text{(iv)} \quad D^n h \geq 0 \text{ on } A \cap (0,N),\ D^n h \leq 0 \text{ on } B \cap (0,N) \text{ and} \\ &\qquad D^n h = 0 \text{ off } (A \cup B) \cap (0,N) \end{aligned}$$

Hence, $h \equiv g$ on $[N,\infty)$. Extend h - g from [0,N] to $W^{n,2}(\mathbb{R})$ by making it N-periodic, calling the resultant function J. Consider

$$F = E + \varepsilon J, \quad \varepsilon > 0.$$

Then $F \in U$ and F is certainly N-periodic since N is even and E has period 2. For $x \in (0,N)$, we have $F^{(n)}(x) = (1-\epsilon)E^{(n)}(x) + \epsilon h^{(n)}(x)$. We note that $h^{(n)}(x)$ has the opposite sign to $E^{(n)}(x)$. Hence, for small ϵ, we have $|F^{(n)}| \leq \alpha(1-\epsilon)$ on $(0,N)$ and so also on $\mathbb{R}$. This contradicts the definition of α in (12.2). From this we conclude that $E^{(n)}$ is a step function with at most 2 discontinuities on any real interval of length 2. In particular, we have established (12.5i).

Now the fact that E is the unique periodic solution implies that $E(x) = E(-x) = E(2i-x)$, $i \in \mathbb{Z}$; so that E is even about the integers and, for n odd, E has integer knots. This establishes the first parts of (12.5v) and (12.5iii). To establish the second part of (12.5v) and hence the second part of (12.5iii) note first that E vanishes at the half-integers. This is a consequence of the fact that E is even about each integer and $E(x) = -E(x-1)$. Thus the function

$$G(x) = E(\tfrac{1}{2} + x) + E(\tfrac{1}{2} - x)$$

vanishes at all integer and half integer points. In particular, on $[0,n]$, G is a spline of degree n with at least 2n zeros (in fact 2n + 1) with at least one zero in each interval $(k,k+n+1)$, $k = -n,\ldots,n-1$. It follows that the spline function G on $[0,n]$ is identically zero by [12.4, p. 524, Lemma 4.2] since such a G can be represented as the linear combination of 2n B-splines of degree n with support in $[0,n]$. Thus E is odd about each half-integer.

It remains to verify (12.5iv) since then (12.5ii) follows. Clearly, we may assume $n \geq 2$. If E' vanishes at a point in $(0,1)$, then on the interval $[0,n-1]$ E' vanishes at 2n - 1 points, viz., the integers $0,\ldots,n-1$ together with n - 1 points interior to the open intervals $(k,k+1)$, $k = 0,\ldots,n-1$. Since E' is a spline of degree $n-1$ on $[0,n-1]$, $E' \equiv 0$ by an argument now familiar. It follows that E' is of one sign on each $(k,k+1)$, $k \in \mathbb{Z}$, and (12.5iv) follows. □

We are now in a position to give an explicit formula for α in terms of the constants K_n of §11.

Corollary 12.5. Let $\alpha = \alpha_n$ be the constant of (12.2) for the data $a_i = (-1)^i$. Then $\alpha = \pi^n/K_n$, $n = 1,2,\ldots$. Here,

$$(12.9) \qquad K_n = \begin{cases} \dfrac{4}{\pi} \displaystyle\sum_{k=0}^{\infty} \frac{1}{(2k+1)^{n+1}} & \text{if } n \text{ is odd} \\[2ex] \dfrac{4}{\pi} \displaystyle\sum_{k=0}^{\infty} (-1)^k \frac{1}{(2k+1)^{n+1}} & \text{if } n \text{ is even.} \end{cases}$$

Proof. Define the functions

$$(12.10) \qquad F_n(x) = \begin{cases} \dfrac{4}{\pi} \displaystyle\sum_{k=0}^{\infty} \frac{\cos(2k+1)x}{(2k+1)^{n+1}} & \text{if } n \text{ is odd,} \\[2ex] \dfrac{4}{\pi} \displaystyle\sum_{k=0}^{\infty} (-1)^k \frac{\cos(2k+1)x}{(2k+1)^{n+1}} & \text{if } n \text{ is even.} \end{cases}$$

These are precisely the functions, which, when properly scaled and normalized, serve as the extremals in the Achieser-Favard-Krein theorem. The F_n are perfect spline functions and have the values $(-1)^i K_n$ at the points πi and $\|F_n^{(n)}\|_\infty = 1$. $F_n^{(n)}$ is a step function with discontinuties at the points πi if n is odd and at the points $\pi(i+\frac{1}{2})$ is n is even. Indeed, if the series (12.10) are differentiated, the Fourier series of these step functions results (cf. [10.5, p. 119]). Since, as we saw in §11, every 2π periodic function f in $W^{n,\infty}(\mathbb{R})$ with mean value zero is uniquely retrievable by convolution of $f^{(n)}$ with the kernel,

$$D_n(x) = \frac{1}{\pi} \sum_{k=1}^{\infty} \frac{\cos(kx-n\pi/2)}{k^n}$$

it follows that E can be expressed in terms of F_n through scaling and normalization; in fact

$$E(x) = \frac{F_n(\pi x)}{K_n} .$$

Since $\|F^{(n)}\|_\infty = 1$, it follows that $\|E^{(n)}\|_\infty = \pi^n/K_n$ and the corollary is proved. □

Remarks. The results of Schoenberg [12.6] which enabled us to conclude that U is nonempty in the course of the proof of Theorem 12.3 are in fact more general and roughly assert that data of power growth can be uniquely interpolated by splines of the same power growth at integer points of the real axis. These and other ideas are set forth in Schoenberg's monograph [12.7]. Theorem 12.4 and Corollary 12.5 reveal that the Euler splines, which are extremals of the Achieser-Favard-Krein theorem, are extremals of the minimum norm problem. In fact, the careful reader may have noted from Corollary 11.2 that the Euler splines are also extremals of the problem of finding best bounding constants for $\|f\|_\infty$, given that f is periodic, $\|f^{(n)}\| \leq 1$, and f has a given number of zero Fourier coefficients. This result was obtained by Favard [12.2] and actually preceded the theorem on trigonometric approximation. In the following chapter, we shall present another extremal problem for which the Euler splines are extremals, viz., the Landau problem which was solved by Kolmogorov. Finally, we remark that Schoenberg's results [12.6] characterize the Euler spline as the unique bounded spline interpolating the data $a_i = (-1)^i$ at the integers i; still another characterization is given by Cavaretta

[12.1].

REFERENCES

12.1 A. Cavaretta, "Perfect splines of minimal sup norm on the real axis," J. Approximation Theory 8 (1973), 285-303.

12.2 J. Favard, "Application de la formule sommatorie d'Euler à la demonstration de quelques propriétés extrémales des integrales des fonctions périodiques ou presque-périodiques," Matematisk Tidsskrift, Ser. B (1936), 81-94.

12.3 S. D. Fisher and J. W. Jerome, "The Euler spline and minimization and interpolation at integer points of the line and half-line," manuscript.

12.4 S. Karlin, Total Positivity, Vol. 1, Stanford University Press, Stanford, California, 1968.

12.5 I. J. Schoenberg, "The elementary cases of Landau's problem of inequalities between derivatives," Amer. Math. Monthly 80 (1973), 121-158.

12.6 _________, "Cardinal interpolation and spline functions, II. Interpolation of data of power growth," J. Approximation Theory 6 (1972), 404-420.

12.7 _________, "Cardinal Spline Interpolation," SIAM, Philadelphia, Pa., 1973.

12.8 J. N. Subbotin, "On the relation between finite differences and the corresponding derivatives, Proc. Steklov Inst. Math. 78 (1965), 24-42. Amer. Math. Soc. Translations (1967).

§13. The Landau Problem and Kolmogorov's Theorem

Let E_n denote the Euler spline derived in Theorem 12.4. Recall that E_n is a perfect spline function of degree n in $W^{n,\infty}(\mathbb{R})$ which is periodic of period two, has maximum modulus one, is strictly monotone on each (k,k+1), has integer knots if n is odd with half-integer knots if n is even, and takes on the values $(-1)^i$ at each integer i. For each $n \geq 1$ and each $\nu = 1,\ldots,n$ define

$$\gamma_{n,\nu} = \max_{x \in \mathbb{R}} |E_n^{(\nu)}(x)| \tag{13.1}$$

and set $\gamma_{n,0} = 1$ for $n \geq 0$. Because of the relation

$$E_n(x) = \frac{F_n(\pi x)}{K_n}, \tag{13.2}$$

where K_n and F_n are defined by (12.9) and (12.10), it follows that the numbers $\gamma_{n,\nu}$ of (13.1) are given by

$$\gamma_{n,\nu} = \pi^{\nu} K_{n-\nu}/K_n, \; n \geq 1, \; 1 \leq \nu \leq n - 1. \tag{13.3}$$

Also, by Corollary 12.5,

$$\gamma_{n,n} = \pi^n/K_n, \; n \geq 1. \tag{13.4}$$

We now state the theorem of Kolmogorov [13.3].

<u>Theorem 13.1</u>. If $f \in W^{n,\infty}(\mathbb{R})$ is such that

$$\|f\|_\infty \leq 1, \; \|f^{(n)}\|_\infty \leq \gamma_{n,n}$$

then

(13.5) $$\|f^{(\nu)}\|_\infty \leq \gamma_{n,\nu} \text{ for } \nu = 1,\dots,n-1.$$

Equivalently, if $M_\nu = \|F^{(\nu)}\|_\infty$, $\nu = 0,1,\dots,n$, then the smallest constant $C_{n,\nu}$ such that

(13.6) $$M_\nu \leq C_{n,\nu} M_0^{1-(\nu/n)} M_n^{\nu/n}, \quad 0 < \nu < n,$$

is given by

(13.7) $$C_{n,\nu} = \gamma_{n,\nu}\gamma_{n,n}^{-\nu/n} = K_{n-\nu}/K_n^{1-\nu/n}, \quad 0 < \nu < n.$$

Proof. (13.5) is clearly implied by (13.6); the converse follows by taking $f(x) = aF(bx)$ and determining a and b so that $\|f\|_\infty = 1$, $\|f^{(n)}\|_\infty = \gamma_{n,n}$. Thus, it suffices to verify (13.5).

We first prove (13.5) in the case when f is periodic of even period k. Let ν be fixed. Note that both f and E_n are of period k. Now define $\beta > 0$ by

(13.8) $$\|f^{(\nu)}\| = \beta\gamma_{n,\nu}.$$

We assume $\beta > 1$ and derive a contradiction. Choose x_0 and x_1 so that

(13.9) $$|f^{(\nu)}(x_0)| = \|f^{(\nu)}\|$$

and

(13.10) $$\beta E_n^{(\nu)}(x_0 - x_1) = f^{(\nu)}(x_0)$$

and define

$$(13.11) \qquad h(x) = E_n(x-x_1) - \frac{1}{\beta} f(x).$$

Then the hypothesis and the fact that $\beta > 1$ imply that h has k distinct zeros because of the equioscillation of E_n. By repeated use of Rolle's theorem (and the periodicity), $h^{(\nu)}$ also has at least k distinct zeros.

We first consider the case $\nu < n - 1$. Then by (13.8), (13.9) and (13.10) x_0 is a local extreme point for both $f^{(\nu)}(x)$ and $E_n^{(\nu)}(x-x_1)$, so

$$(13.12) \qquad f^{(\nu+1)}(x_0) = E_n^{(\nu+1)}(x_0-x_1) = 0;$$

hence $h^{(\nu)}(x_0) = h^{(\nu+1)}(x_0) = 0$. So on taking to account the k zeros of $h^{(\nu)}(x)$ and the double zero of $h^{(\nu)}(x)$ at x_0, we conclude that $h^{(\nu+1)}$ has at least k + 1 distinct zeros. It follows by Rolle's theorem that

$$(13.13) \qquad h^{(n)}(x) = E_n^{(n)}(x-x_1) - (1/\beta)f^{(n)}(x)$$

must have at least k + 1 sign changes. But by the hypothesis and $1/\beta < 1$,

$$(13.14) \qquad \left|\frac{1}{\beta} f^{(n)}(x)\right| < |E_n^{(n)}(x-x_1)|$$

and so $h^{(n)}(x)$ has exactly k sign changes. Hence a contradiction and so $\beta \leq 1$, which proves the theorem by (13.8).

For the case $\nu = n - 1$, we observe that by the hypothesis and $1/\beta < 1$, h(x) exhibits k zeros where the function actually changes sign; hence by Rolle's theorem $h^{(n-1)}(x)$ also has k zeros where $h^{(n-1)}(x)$ changes sign. Moreover, from (13.10)

$h^{(n-1)}(x_0) = 0$. But x_0 is also a local extreme point of $h^{(n-1)}(x)$ since by (13.8), (13.9) and (13.10) we see that $E_n^{(n)}(x-x_1)$ must change sign at x_0. Therefore $h^{(n-1)}(x)$ has at least $k + 1$ distinct zeros. Hence $h^{(n)}(x)$ has at least $k + 1$ changes of sign, and this is a contradiction, as before.

We have thus proved the normalized Kolmogorov theorem under the restriction that f have integral period. We now prove the general result using this special case. Let f be an arbitrary function in our class and as before set

$$M_\nu = M_\nu(f) = \|f^{(\nu)}\|, \quad \nu = 0,\dots,n. \tag{13.15}$$

We assume the hypothesis and prove (13.5). Now we associate with f a periodic function F in such a way that $M_\nu(F)$ will be close to $M_\nu(f)$. To do this, we need the following auxiliary function:

$$g(x) = \begin{cases} 1 & -1 \le x \le 1 \\ (-1)^n(x-2)^n \sum_{k=0}^{n-1} \binom{n+k-1}{k}(x-1)^k, & 1 < x < 2 \\ (-1)^n(x+2)^n \sum_{k=0}^{n-1} \binom{n+k-1}{k}(x+1)^k, & -2 < x < -1 \\ 0 & |x| \ge 2. \end{cases}$$

By construction g is in our class and has bounded derivatives up to order n. Now let k be a positive integer and define $F_k(x)$ by the relations

$$F_k(x) = f(x)g\left(\frac{x}{k}\right), \quad -2k \le x \le 2k \tag{13.16}$$

and

$$(13.17) \qquad F_k(x+4k) = F_k(x) \text{ for all } x.$$

Thus $F_k(x)$ has period $4k$ and $F_k(x)$ is in our class since $g(x/k)$ has zeros of multiplicity n at $\pm 2k$.

To compute $M_\nu(F_k)$ we apply the Leibniz formula to the product $f(x)g(x/k)$. According to our first argument in this section of the proof and the construction of g, we can choose a constant A such that

$$(13.18) \qquad \begin{aligned} M_\nu(f) &< A, \quad \nu = 0,\dots,n, \\ M_\nu(g) &< A, \quad \nu = 0,\dots,n. \end{aligned}$$

Then for any $\nu = 0,\dots,n$, and $-2k \le x < 2k$,

$$(13.19) \qquad (F_k(x))^{(\nu)} = (f(x)g(x/k))^{(\nu)}$$

$$= f^{(\nu)}(x)g\left(\frac{x}{k}\right) + \sum_{i=1}^{\nu} \binom{\nu}{i} f^{(\nu-i)}(x) \frac{d^i}{dx^i} g\left(\frac{x}{k}\right).$$

We observe that for each x

$$(13.20) \qquad g(x/k) \to 1 \text{ as } k \to \infty,$$

and so

$$(13.21) \qquad f^{(\nu)}(x)g(x/k) \to f^{(\nu)}(x) \text{ as } k \to \infty.$$

As for the finite sum appearing in (13.19), we have that

$$(13.22) \qquad \frac{d^i}{dx^i} g\left(\frac{x}{k}\right) = \frac{1}{k^i} g^{(i)}\left(\frac{x}{k}\right), \quad (i = 1,\dots,\nu),$$

and so

$$(13.23)\qquad \left| \sum_{i=1}^{\nu} \binom{\nu}{i} f^{(\nu-i)}(x) \frac{d^i}{dx^i} g\left(\frac{x}{k}\right) \right| \leq \frac{1}{k} \cdot 2^{\nu} \cdot A^2.$$

Hence from (13.19), (13.20), and (13.23) we conclude that

$$(13.24)\qquad \lim_{k \to \infty} M_{\nu}(F_k) = M_{\nu}(f), \ \nu = 0,\ldots,n.$$

Now the desired result is immediate. Let $\beta < 1$ and consider βf. Then

$$(13.25)\qquad \|\beta f\| \leq \beta < 1$$

and

$$(13.26)\qquad \|\beta f^{(n)}\| \leq \beta \gamma_{n,n} < \gamma_{n,n}.$$

Now

$$(13.27)\qquad \lim_{k \to \infty} M_{\nu}(\beta F_k) = M_{\nu}(\beta f), \ \nu = 0,1,\ldots,n$$

and hence, for sufficiently large k, we have

$$(13.28)\qquad M_0(\beta F_k) \leq 1$$

and

$$(13.29)\qquad M_n(\beta F_k) \leq \gamma_{n,n}.$$

So by the result for periodic functions already proved,

$$M_\nu(\beta F_k) \leq \gamma_{n,\nu}, \quad (\nu = 1,\ldots,n-1),$$

whence it follows by (13.27) that

$$M_\nu(\beta f) \leq \gamma_{n,\nu}, \quad (\nu = 1,\ldots,n-1). \tag{13.30}$$

Letting β tend to 1, we conclude that

$$M_\nu(f) \leq \gamma_{n,\nu}, \quad (\nu = 1,\ldots,n-1). \tag{13.31}$$

Remarks. The proof of Theorem 13.1 which we have presented is due to Cavaretta [13.1]. The result was originally obtained by Komogorov in 1939 [13.3]. In his paper, Kolmogorov notes that the Euler splines E_n are extremals of (13.6); in fact, Kolmogorov was aware of their use by Achieser and Krein [11.1] and comments to this effect. Kolmogorov also notes that, for any extremal F of (13.6), the function $G(x) = aF(bx+c)$, $a > 0$, $b > 0$, is also an extremal. The formulation given by (13.5) is due to Cavaretta [13.1] and Schoenberg [12.5]; its extremals are uniquely given by the Euler splines.

The problem was originally posed by Landau [13.4] and solved by him for the case $n = 2$, the case of lowest order. Others also worked on the problem prior to the complete solution given by Kolmogorov, e.g., Hadamard and Shilov (cf. the bibliography of [13.3]). An analysis of the Landau problem on the half line has been given by Cavaretta and Schoenberg [13.5] and results for the corresponding problem on a finite interval have recently been announced by Karlin [13.2].

REFERENCES

13.1 A. S. Cavaretta, "An elementary proof of Kolmogorov's theorem," Amer. Math. Monthly, 81 (1974), 480-486.

13.2 S. Karlin, "Some variational problems on certain Sobolev spaces and perfect splines," BAMS, 79 (1973), 124-128.

13.3 A. N. Kolmogorov, "On inequalities between the upper bounds of the successive derivatives of an arbitrary function on an infinite interval," Amer. Math. Soc. Trans. Series 1, 2 (1962), 233-243 (Russian, 1939).

13.4 E. Landau "Einige Ungleichungen für zweimal differenzierbare Funktionen," Proc. London Math. Soc., (2), 13 (1913), 43-49.

13.5 I. J. Schoenberg and A. S. Cavaretta, "Solution of Landau's problem concerning higher derivatives on the half line," MRC Technical Summary Report 1050, Madison, Wisconsin, 1970.

§14. Perfect Interpolating Splines on Compact Intervals

In §7, we presented a result concerning multipoint interpolation with minimal norm bang-bang properties for derivatives. In fact, the results are extremely general and hold for inequality as well as equality constraints. In this chapter, we shall show that under very special equality constraints, one can give a precise global bound on the number of knots of the perfect spline, which sharpens the local bounds presented in §7. The theorem we shall present is due to S. Karlin [13.2], but this proof is an adaptation of one of C. DeBoor [14.1].

Theorem 14.1. Let $f_0 \in W^{n,\infty}(a,b)$, $n \geq 1$, and let distinct points $a \leq x_1 < \dots < x_m \leq b$ be given, together with positive integers $1 \leq k_i \leq n$, $i = 1,\dots,m$. Let f_0 be a prescribed element of $W^{n,\infty}(a,b)$ and let $U = \{f \in W^{n,\infty}(a,b) : f^{(j)}(x_i) = f_0^{(j)}(x_i),$ $0 \leq j \leq k_i - 1,\ 1 \leq i \leq m\}$. If $\sum_{i=1}^{m} k_i = n + r$, $r \geq 0$, then there exists a perfect spline function $s \in U$ of degree n with less than r interior knots on (a,b) satisfying

$$\|s^{(n)}\|_\infty = \inf_{f \in U} \|f^{(n)}\|_\infty . \tag{14.1}$$

Proof. Denote by S the r dimensional linear space of spline functions of degree n - 1 spanned by the B-splines $\{M_i(x)\}_{i=1}^{r}$; here the B-splines are the nth. order (confluent) divided differences of $n(\cdot - x)_+^{n-1}$ taken on n + 1 consecutive points of the x_i including multiplicities. Then the class $U^{(n)}$ may be written explicitly in terms of $M_1,\dots,M_r$ by

$$U^{(n)} = \{f^{(n)} : f \in W^{n,\infty}(a,b) \text{ and } \int_a^b M_i[f^{(n)} - f_0^{(n)}] = 0,\ i = 1,\dots,r\}$$

since these relations assert that the nth. order divided

differences of f and f_0 agree [11.3]. Thus we may conveniently work with $U^{(n)} \subset L^\infty(a,b)$, consisting of nth. order derivatives of members of U. There is an element $f_*^{(n)} \in U^{(n)}$ of minimal $L^\infty(a,b)$ norm; this follows from the Hahn-Banach theorem applied to the functional on $S \subset L^1(a,b)$ determined by $f_0^{(n)}$. Now for each $\epsilon > 0$, define the smoothed space $S_\epsilon \subset L^1(a,b)$ by

$$S_\epsilon = K_\epsilon S = \{\varphi(x) = \frac{1}{\epsilon\sqrt{2\pi}} \int_{-\infty}^{\infty} \exp(-(x-\xi)^2/2\epsilon^2) g(\xi) d\xi, \ g \in S\}.$$

Note that, although the members of S_ϵ are defined on $\mathbb{R}$, we restrict attention to [a,b]. The function $f_*^{(n)}$ defines a linear functional λ_ϵ on S_ϵ in the usual way:

$$\lambda_\epsilon \varphi = \int_a^b \varphi(\xi) f_*^{(n)}(\xi) d\xi, \ \varphi \in S_\epsilon.$$

Since S_ϵ is finite dimensional, there is an element $\varphi_\epsilon \in S_\epsilon$ such that,

$$(14.2) \qquad \lambda_\epsilon \varphi_\epsilon = \sup\{\lambda_\epsilon \varphi : \varphi \in S_\epsilon, \ \|\varphi\|_1 = 1\},$$

i.e., $\|\lambda_\epsilon\| = \lambda_\epsilon \varphi_\epsilon$. By the Hahn-Banach theorem λ_ϵ may be extended to $L^1(a,b)$ in a norm-preserving fashion; thus, since $L^\infty(a,b)$ is the dual of $L^1(a,b)$, there is a function $h_\epsilon \in L^\infty(a,b)$ satisfying

$$(14.3) \qquad \|h_\epsilon\|_\infty = \lambda_\epsilon \varphi_\epsilon.$$

We claim that such an h_ϵ may be given by

$$(14.4) \qquad h_\epsilon = \lambda_\epsilon \varphi_\epsilon \ \text{signum}\ \varphi_\epsilon.$$

Indeed, the smoothing operation ensures that φ_ε cannot vanish on a set of measure zero; in fact, φ_ε has at most r - 1 zeros [12.4, p. 528, eq. (4.23)]. Thus, for fixed $\psi \in S_\varepsilon$, the real-valued function of t,

$$\Phi_\varepsilon(t) = \frac{\lambda_\varepsilon \varphi_\varepsilon + t\lambda_\varepsilon \psi}{\|\varphi_\varepsilon + t\psi\|_1},$$

is differentiable in a neighborhood of zero and clearly satisfies $\Phi_\varepsilon'(0) = 0$ by (14.2). Since [14.3]

$$\frac{d}{dt}\left(\|\varphi_\varepsilon + t\psi\|\right)_{t=0} = \int_a^b \psi \text{ signum } \varphi_\varepsilon,$$

it follows that

$$\lambda_\varepsilon \psi = \int_a^b [\lambda_\varepsilon \varphi_\varepsilon \text{ signum } \varphi_\varepsilon]\psi$$

for each $\psi \in S_\varepsilon$ and since the function given by (14.4) satisfies (14.3) the claim is established. Now the functions (14.4) have the property that they are bounded in $L^\infty(a,b)$; in fact,

$$\|h_\varepsilon\| = \lambda_\varepsilon \varphi_\varepsilon = \|\lambda_\varepsilon\| \leq \|f_*^{(n)}\|_\infty.$$

Thus, since the ball of radius $\|f_*^{(n)}\|_\infty$ is weak-$*$ sequentially compact in $L^\infty(a,b)$ it follows that there is a function $h \in L^\infty(a,b)$ and a sequence $\varepsilon_\nu \to 0$ such that h_{ε_ν} converges to h in the weak-$*$ topology. By taking subsequences, if necessary, we may assume that there are points $a \leq \xi_1 < \ldots < \xi_{k-1} \leq b$ such that $k \leq r$ and such that the knot sequences of h_{ε_ν} converge to $\xi_1, \ldots, \xi_{k-1}$ as $\varepsilon_\nu \to 0$. It follows that h is a step function with no more than k - 1 interior discontinuities on (a,b) satisfying

$$(14.5)\qquad |h| = \|h\|_\infty \leq \liminf_{\nu\to\infty} \|h_{\varepsilon_\nu}\| \leq \|f_*\|_\infty .$$

Moreover, $h \in U^{(n)}$ since the smoothing relation defines an approximate identity K_ε such that $\|\varphi - K_\varepsilon\varphi\|_1 \to 0$ as $\varepsilon \to 0$ for each $\varphi \in S$. Specifically, for each $\varphi \in S$,

$$\int_a^b h\varphi - \int_a^b f_*^{(n)}\varphi = \int_a^b (h-h_{\varepsilon_\nu})\varphi + \int_a^b h_{\varepsilon_\nu}(\varphi - K_{\varepsilon_\nu}\varphi) + \int_a^b f_*^{(n)}(K_{\varepsilon_\nu}\varphi - \varphi)$$

and the first of these three sequences converges to zero by the weak-* convergence of h_{ε_ν} to h, while the second and third tend to zero by the L^1 convergence of $K_{\varepsilon_\nu}\varphi$ to φ (note that $\|h_{\varepsilon_\nu}\|_\infty$ is bounded). It follows from (14.5) that $\|h\|_\infty = \|f_*^{(n)}\|_\infty$ and the proof is complete.

Remarks. We have presented here our adaptation of DeBoor's proof [14.1] of Karlin's theorem. In so doing we have given the details of the fact that $h \in U^{(n)}$ and corrected a minor error in [14.1], viz., the functionals λ_ε must be defined by $f_*^{(n)}$ rather than $f_0^{(n)}$ in order to imply that h is of minimal norm in $U^{(n)}$. It is possible to generalize this result to more general differential operators and more general interpolation conditions provided the analogue of the B-splines is present. The reader is referred to Karlin's book [12.4] for an exposition of Chebyshevian B-splines. Even more general B-splines have been constructed by Jerome and Schumaker [14.2]. We shall not pursue the evident generalizations of Theorem 14.1 via such B-splines.

REFERENCES

14.1 C. DeBoor, "A remark concerning perfect splines," Bull. Amer. Math. Soc., 80 (1974), 724-727.

14.2 J. Jerome and L. Schumaker, "Local support bases for a class of spline functions," M.R.C. Technical Summary Report 1255, Madison, Wisconsin, 1974.

14.3 J. Rice, The Approximation of Functions, vol. I, Addison-Wesley, Reading, Mass., 1964.

PART VIII. FURTHER TOPICS

§15. A Pólya Algorithm for the Favard Solution, N-Width Characterizations and Whitney Type Theorems

15.1 The Favard Solution

Let $[a,b]$ be a given interval, let $1 < p \leq \infty$ and let $\varphi_1,\ldots,\varphi_n$ be linearly independent elements of $L^q(a,b)$, $\frac{1}{p} + \frac{1}{q} = 1$ and let $r_1,\ldots,r_n$ be given real numbers. Consider the minimization problem

$$(15.1) \qquad \alpha_p = \inf\{\|g\|_p : \int_a^b g\varphi_j = r_j,\ 1 \leq j \leq n\}.$$

As emerged from §14, this problem is equivalent, for example, to minimizing the L^∞ norm of the kth. derivatives of $W^{k,\infty}(a,b)$ functions which interpolate certain prescribed values at the $n + k$ points of a mesh with multiplicity not exceeding k; in this case, the functions φ_i are the (confluent) B-splines and the r_i the kth. order divided differences of the given interpolation values. In this section, we present C. DeBoor's account [15.1] of an older result due to J. Favard [15.4] on the construction of a solution of (15.1) which is in some sense optimal. The reader will recall that the process described in §6 shows that every solution to the minimization problem (15.1) is determined uniquely on some set of positive measure related to the functions $\varphi_1,\ldots,\varphi_n$, but we have no information available in general on where the function h, defined from Theorem 3.3, vanishes. As this set may have positive measure, there may be and indeed are (infinitely) many solutions of (15.1) which differ where $h = 0$. Favard's procedure produces in a finite number of steps (no more than n) a solution g to (15.1) such that if g_1 is any other solution and $|g_1| \leq |g|$ a.e. then $g_1 = g$ on $[a,b]$. The function g will be referred to as

"Favard's solution." It has the feature that it is the limit in $L^1(a,b)$, as $p \to \infty$, of the L^p extremal solutions. This will be discussed in the next section and provides a result of Polya type, comparable to the situation in the discrete ℓ^p spaces.

For the remainder of this section we fix $p = \infty$ and select a function $g_0 \in L^\infty(a,b)$ satisfying $\int_a^b g_0\varphi_j = r_j$, $j = 1,\ldots,n$. Here, of course, $\varphi_1,\ldots,\varphi_n$ are $L^1(a,b)$ functions whose linear span is denoted by S. DeBoor describes Favard's procedure as follows.

Step 1. Set $N_1 = [a,b]$ and set $i = 1$.

Step 2. Set $G_i = \{g \in L^\infty(N_i) : \int_{N_i} g\varphi = \int_{N_i} g_0\varphi \text{ for all } \varphi \in S\}$ and set

$$m_i = \inf\{\|g\|_{L^\infty(N_i)} : g \in G_i\}$$

and pick a function $g_i \in G_i$ at which the infimum is attained.

Step 3. Let $S_i = \{\varphi|N_i : \varphi \in S\}$. If $\dim S_i = 0$, then set $N_{i+1} = \emptyset$. Otherwise, choose a function $h_i \in S$ such that $h_i|N_i$ is an L^1-extremal for the linear functional λ_i given by

$$\lambda_i(\varphi) = \int_{N_i} g_0\varphi, \ \varphi \in S_i$$

and set $N_{i+1} = \{t \in N_i : h_i(t) = 0\}$.

Step 4. Redefine g_0 to be equal to g_i on N_i

Step 5. If N_{i+1} has positive measure, increase i by 1 and go to step 2. Otherwise, stop.

The process clearly ends in n or fewer steps since $\dim S_{i+1} < \dim S_i$ for each i; indeed the linear map which restricts each element of

S_i to N_{i+1} has a non-trivial kernel.

Theorem 15.1. The foregoing procedure produces a function $g^* \in G_1$ with $\|g^*\|_\infty = \inf\{\|g\|_\infty : g \in G_1\}$ and decreasing sequences $N_1 \supseteq \dots \supseteq N_r$ and $m_1 \geq \dots \geq m_r \geq 0$ for some $r \leq n$ so that

(i) $|g^*| = m_i$ on $N_i \setminus N_{i+1}$;

(ii) if $g \in G_1$ and $|g| \leq |g^*|$ on $N_1 \setminus N_{i+1}$, then $g = g^*$ on $N_1 \setminus N_{i+1}$ for $i = 1,\dots,r$; in particular, if $|g| \leq |g^*|$ on $[a,b]$ then $g = g^*$;

(iii) g^* depends only on $\varphi_1,\dots,\varphi_n$ and the numbers $r_1,\dots,r_n$.

Proof. If we define g^* to be the function produced by the Favard procedure then clearly g^* satisfies (i) and (ii). We need only show that (iii) holds and this is proved by induction on dim S. If dim S = 0 then the assertion is trivial so assume that dim S > 0. If g' is another function produced by Favard's procedure with associated sequences $N_1 \supseteq N_2' \supseteq \dots \supseteq N_{r'}'$, $m_1' \geq m_2' \geq \dots \geq m_{r'}' \geq 0$ and $h_1',\dots,h_{r'}'$ and g' satisfies (i) and (ii) above with N_i' and m_i' replacing N_i and m_i, then $|g'| \leq |g^*|$ on $N_1 \setminus N_j$ where j is the smallest integer such that $m_1 > m_j$. Hence, $g' = g^*$ on $N_1 \setminus N_j$. If j' is the smallest integer such that $m_1' > m_{j'}'$, then clearly $N_{j'}' \subseteq N_j$, and so $N_j = N_{j'}'$ by symmetry. Hence,

$$\int_{N_j} g^* \varphi = \int_{N_{j'}'} g' \varphi, \quad \varphi \in S.$$

Hence, the induction hypothesis may be applied on the set N_j since $\dim S_j < \dim S$ and we find $g' = g^*$ a.e. on $[a,b]$.

15.2. The Convergence of L^p Solutions to Favard's Solution

Let $1 < p < \infty$ and choose a function $w \in L^p(a,b)$ for which $\int_a^b w\varphi_j = r_j$, $j = 1,\ldots,n$. Here, $\varphi_1,\ldots,\varphi_n$ are $L^q(a,b)$ functions, as noted, with linear span S. Let ψ_q be the unique solution to the L^q extremal problem

$$(15.2) \qquad \alpha_q' = \sup\{\int_a^b \psi w : \psi \in S, \ \|\psi\|_q = 1\}.$$

A standard duality argument, such as that of Theorem 3.1, shows that, if g_p is the unique solution of the minimization problem (15.1) for $1 < p < \infty$, then $\alpha_p = \alpha_q'$ and

$$(15.3) \qquad g_p = \alpha_p |\psi_q|^{q-1} \operatorname{sign} \psi_q, \quad \frac{1}{p} + \frac{1}{q} = 1.$$

We present now a result of Pólya type due to Chui, Smith and Ward [15.2].

<u>Theorem 15.2</u>. The net $\{g_p\}$, $1 < p < \infty$, is convergent in $L^1(a,b)$ to the Favard solution described in Theorem 15.1, i.e., $g_p \to g^*$ as $p \to \infty$. Here, g_p is the unique solution of (15.1) and $w,\varphi_1,\ldots,\varphi_n$ are assumed to be in $L^\infty(a,b)$.

<u>Proof</u>. The method of proof is a standard one. For any sequence $p_\nu \to \infty$ we construct a subsequence $p_{\nu'}$ such that $g_{p_{\nu'}} \to g_*$ in $L^1(a,b)$. Now Chui, Smith, and Ward describe what they call the L^p algorithm as follows. Let $p_\nu \to \infty$.

<u>Step 0</u>. Set $\Sigma_1 = [a,b]$, $M_1 = S$ and $i = 1$.

<u>Step 1</u>. Let $\sigma_i = \lim_{q_\nu \to 1} \|\psi_{q_\nu}\|_{L^\infty(\Sigma_i)}^{q_\nu - 1} \leq 1$, after passing to an appropriate subsequence if necessary.

Step 2. If $\sigma_i = 0$, set $s = 0$ on Σ_i and stop. Otherwise, let

$$\psi^i = \lim_{q_\nu \to 1} (\psi_{q_\nu}|\Sigma_i)(\|\psi_{q_\nu}\|_{L^\infty(\Sigma_i)})^{-1}$$

again passing to an appropriate subsequence if necessary. Here the limit may be taken in any norm on the finite dimensional space S and we may assume, without loss of generality, pointwise convergence of ψ_{q_ν} a.e. on Σ_i.

Step 3. Let $\Sigma_{i+1} = \{t \in \Sigma_i : \psi^i(t) = 0\}$ and set $s = \alpha_\infty \sigma_i$ sign ψ^i on $\Sigma_i \backslash \Sigma_{i+1}$.

Step 4. Let M_{i+1} be the restriction of M_i to Σ_{i+1}. If $\dim M_{i+1} = 0$, then stop. Otherwise, increase i by 1 and return to Step 1.

We note that $\dim M_1 = n$ and $\dim M_{i+1} < \dim M_i$ so that steps 1 and 2 can be carried out; indeed, the sequence ψ_{q_ν} is bounded in $L^1(a,b)$ by Hölder's inequality and hence in $L^\infty(a,b)$ by the equivalence of norms on S. Moreover, the entire process ends after no more than n steps. We also note the fact that

$$\alpha_p \to \alpha_\infty, \text{ as } p \to \infty,$$

which is a consequence of the characterization (15.2) of α_p $(1 < p \leq \infty)$ together with the well known fact that $\|\varphi\|_q \to \|\varphi\|_1$ as $q \to 1$ for a function $\varphi \in L^\infty(a,b)$, the latter space being dense in $L^q(a,b)$, $q < \infty$.

Now at any point (up to a set of measure zero) at which $\psi^i \neq 0$ we have $g_{p_\nu} \to \alpha_\infty \sigma_i$ sign(ψ^i) as $p_\nu \to \infty$ by the formula (15.3). In making this observation, we use the representation

$$|\psi_{q_\nu}|^{q_\nu-1} = \|\psi_{q_\nu}\|^{q_\nu-1}_{L^\infty(\Sigma_i)} \left(\frac{|q_\nu|}{\|q_\nu\|_{L^\infty(\Sigma_i)}} \right)^{q_\nu-1}$$

together with steps 1 and 2 and the fact that $q_\nu \to 1$ as $p_\nu \to \infty$. Now, from (15.3), we conclude from Hölder's inequality that the functions g_p are bounded in $L^1(a,b)$ and hence in $L^\infty(a,b)$ as before. Thus, by the dominated convergence theorem and the earlier discussion we have $g_{p_\nu} \to s$ in $L^1(a,b)$ where

$$(15.4) \qquad s = \begin{cases} \alpha_\infty \sigma_i \text{ sign } \psi^i & \text{on } \Sigma_{i+1} \backslash \Sigma_i \\ 0 & \text{on } \Sigma_{r+1} \end{cases}$$

where r is the smallest integer for which $\dim M_{r+1} = 0$ or for which $\sigma_{r+1} = 0$.

We shall now demonstrate that $s = g^*$. First note that s is a solution of (15.1) for $p = \infty$; indeed, $\|s\|_\infty \leq \alpha_\infty$ by (15.4) and the $L^1(a,b)$ convergence of g_{p_ν} to s, together with the relations

$$r_j = \int_a^b g_{p_\nu} \varphi_j, \; j = 1,\ldots,n,$$

imply

$$r_j = \int_a^b s\varphi_j, \; j = 1,\ldots,n.$$

Here we have used the fact that $\varphi_1,\ldots,\varphi_n \in L^\infty(a,b)$. Begin Favard's procedure with $g_0 = s$ and note that at the first stage we may choose $m_1 = \alpha_\infty$ and $h_1 = \tau_1 \psi^1$. Here, $\tau_1 = \lim_{q_\nu \to 1} \|\psi_{q_\nu}\|_\infty$, where a subsequence has been chosen, if necessary, to insure convergence. Thus,

$$\alpha_\infty = \lim_{q_\nu \to 1} \int_a^b \psi_{q_\nu} w = \tau_1 \int_a^b \psi^1 w = \int_a^b h_1 w$$

and clearly we may assume $\tau_1 \neq 0$ otherwise $\alpha_\infty = 0$ and the problem is trivial. We conclude that $N_2 = \Sigma_2$ and s is thus unchanged on $N_1 \backslash N_2$ by the Favard procedure. Inductively, suppose we have reached the ith. stage of Favard's procedure with s unchanged on $N_1 \backslash N_{i+1}$. If $g \in G_{i+1}$, in the terminology of step 2 of the previous section, then

$$\int_{N_{i+1}} s\psi^{i+1} = \int_{N_{i+1}} g\psi^{i+1}$$

and we deduce that $s|N_{i+1}$ is of minimum L^∞-norm from the relations

$$\begin{aligned} \|s\|_{L^\infty(N_{i+1})} &= \left(\int_{N_{i+1}} s\psi^{i+1}\right)\left(\|\psi^{i+1}\|_{L^1(N_{i+1})}\right)^{-1} \\ &= \left(\int_{N_{i+1}} g\psi^{i+1}\right)\left(\|\psi^{i+1}\|_{L^1(N_{i+1})}\right)^{-1} \\ &\leq \sup\{|\int_{N_{i+1}} g\varphi| \; : \; \|\varphi\|_{L^1(N_{i+1})} = 1\} \\ &= \|g\|_{L^\infty(N_{i+1})}. \end{aligned}$$

Thus s is not changed at this step. Moreover, if the Favard procedure has not terminated, then h_{i+1} is a constant (nonzero) multiple of ψ^{i+1} and it follows that $\Sigma_{i+2} = N_{i+2}$ and the induction is completed.

15.3 N-Widths and Perfect Spline Extremals.

In this section we shall state without proof some remarkable results of Tihomirov on the computation of the N-widths of the classes W_n^* and W_n, defined in Chapter Eleven. By way of recall, we note that W_n is the preimage in $W^{n,\infty}(-1,1)$ of the unit ball of $L^\infty(-1,1)$ under D^n. W_n^* is the class of those 2π periodic functions which is the preimage under D^n of the unit ball intersected with the flat of mean value zero functions in $L^\infty(-\pi,\pi)$. The N-widths are stated for spaces of continuous functions and periodic continuous functions.

The notion of N-width was exploited in Chapter Ten when certain convergence rates were shown to be best possible in an asymptotic sense. For the case $p = \infty$, we essentially used the known result that $d_N(W_n) = O(N^{-n})$, $N \to \infty$. It is precisely this result which we shall now investigate in some detail.

The notion of N-width was introduced by Kolmogorov [15.9] as a means of characterizing the approximability of a given set $\mathscr{A}$ in a normed linear space X by linear spaces of dimension N. Indeed, we take the maximum distance

$$(15.5) \qquad E(\mathscr{A},M) = \sup_{f \in \mathscr{A}} \inf_{g \in M} \|f-g\|$$

of $\mathscr{A}$ from an N dimensional subspace M of X and then vary the subspace and take the infimum

$$(15.6) \qquad d_N(\mathscr{A}) = \inf\{E(\mathscr{A},M) : M \subset X, \dim M = N\}$$

of the distances for all possible subspaces. Tihomirov, in a series of papers [15.14, 15.15, 15.16, 15.17] has calculated $d_N(W_n)$ in $C[-1,1]$ and $d_N(W_N^*)$ in $C^*[-\pi,\pi]$ both in an asymptotic sense, and also, via certain explicit characterizations.

<u>Theorem 15.3</u>. The N-widths of W_n^* in $C^*[-\pi,\pi]$ are given explicitly by

$$(15.7)\quad d_0(W_n^*) = \infty,\ d_{2N-1}(W_n^*) = d_{2N}(W_n^*) = K_n/N^n,\ N = 1,2,\ldots$$

where K_n is defined in Theorem 11.1. Moreover, if L_{2N} denotes the 2N - 1 dimensional space of periodic spline functions of degree n with knots on the interval $[0,2\pi]$ at $k\pi/N$, $k = 1,\ldots,2N$, then

$$(15.8)\qquad d_{2N-1}(W_n^*) = d_{2N}(W_n^*) = E(W_n^*, L_{2N}),\ N = 1,2,\ldots,$$

where $E(\cdot,\cdot)$ is defined by (15.5). Moreover, the space T_{N-1}, of trigonometric polynomials of degree N - 1, is a second extremal subspace of dimension 2N - 1, N = 1,2,...; specifically,

$$(15.9)\qquad d_{2N-1}(W_n^*) = E(W_n^*, T_{N-1}),\ N = 1,2,\ldots\ .$$

This theorem is established in [15.17], althought the special case (15.9) was known much earlier to Tihomirov [15.15]. An intriguing aspect of this result is that the N-widths (15.7) are expressible as norms of perfect spline functions

$$f_{nN}(x) = (-1)^{[N/2]} \frac{4}{\pi N^n} \sum_{k=0}^{\infty} (-1)^{k(n+1)} \frac{\cos N(2k+1)x}{(2k+1)^{n+1}}$$

whose knots are used to define extremal spline subspaces (cf. (15.8)). The functions f_{nN}, with N = m - 1, were precisely the extremals of Theorem 11.1 and are related to the Euler spline as explained in Chapter 12. It is reasonable therefore to conjecture the existence of a perfect spline function with irregularly spaced knots which plays the same role in the determination of $d_N(W_n)$. This has been established by Tihomirov [15.16, 15.17] in one of the deeper theorems presented here.

<u>Theorem 15.4</u>. Let m and n be integers satisfying $m \geq 0$, $n \geq 1$. The minimization problem

$$\lambda_{mn} = \inf \|x\|_{L^\infty(-1,1)} \tag{15.10}$$

subject to $x \in W^{n,\infty}(-1,1)$ and the conditions

$$\begin{aligned} &\text{(i)} \quad x^{(n+1)} = 2 \sum_{k=1}^{\nu} (-1)^{k+1} \delta(t-t_k) \\ &\text{(ii)} \quad -1 \leq t_1 \leq \cdots \leq t_\nu \leq 1, \ \nu \leq m, \\ &\text{(iii)} \quad x^{(n)}(t) \equiv -1, \ t \leq -1, \end{aligned} \tag{15.11}$$

has a solution $x_{m,n}$ satisfying

(15.12)
(i) x_{mn} assumes its maximum and minimum values (successively alternately) at $m + n + 1$ points of $[-1,1]$;
(ii) $x_{mn}^{(n)}$ has modulus one and has exactly m changes of sign in $(-1,1)$.

Moreover, the N-widths of W_n satisfy

$$d_N(W_n) = \begin{cases} \infty & \text{if } N \leq n-1, \\ \|x_{N-n,n}\| & \text{if } N \geq n \end{cases} \tag{15.13}$$

and, if $L_{N-n,n}$ denotes the N dimensional space of spline functions of degree n with knots at the N - n break points of $x^{(n)}_{N-n,n}$, we have

$$d_N(W_n) = E(W_n, L_{N-n,n}),\ N = n,\ n+1, \ldots .$$

15.4. Theorems of Whitney Type

Let E be a closed subset of [a,b], f a real-valued function defined on E and n a positive integer. In 1934, H. Whitney [15.18] obtained a necessary and sufficient condition, in terms of divided differences, for the existence of a function $F \in C^n[a,b]$ such that $F(x) = f(x)$, $x \in E$. Specifically, the condition requires that the nth. order divided differences $f(x_0, \ldots, x_n)$ converge to a limit as $x_0, \ldots, x_n$ tend to an arbitrary accumulation point in E for distinct points $x_0, \ldots, x_n$ in E. Whitney also showed that the same condition is necessary and sufficient for the C^n-extension of a Taylor field $f_0(x), \ldots, f_{\mu(x)-1}$ of variable height on E; in this case a point $x_0 \in E$ assumes multiplicity $\mu(x_0)$ and the divided differences of order $k \leq \mu(x_0) - 1$ are replaced by $f_k(x_0)$.

In this section we shall present comparable results characterizing the possibility of extension of functions defined on closed subsets of $\mathbb{R}$ to the spaces $L^{n,p}(\mathbb{R})$, $1 < p < \infty$, where these spaces represent n-fold integrals of $L^p(\mathbb{R})$ functions. We shall present these results without proof for which the reader is referred to the paper of Golomb [15.6] which serves as our guide for this section. As noted in [15.6], the theorem of Whitney is in fact valid for $E \subset \mathbb{R}$ so that the problem of this section is a parallel problem.

Now the special case when $E = \mathbb{R}$ deserves special mention. In this case, the criteria characterize $L^{n,p}(\mathbb{R})$ functions in terms of divided differences. In this connection we have

Theorem 15.5. The real-valued function f is in $L^{n,p}(\mathbb{R})$ if and only if

$$\sup \sum_{i=1}^{N-n} |f(x_i,\ldots,x_{i+n})|^p (x_{i+n}-x_i) < \infty$$

for arbitrary N-tuples $x_1 < \ldots < x_N$ in $\mathbb{R}$, $N \geq n+1$.

Theorem 15.5 was obtained in the case $n = 1$, $1 < p < \infty$ by F. Riesz [15.10], in the case $n \geq 1$, $p = 2$ by Schoenberg [15.11] and in the general case by Jerome and Schumaker [3.5]. The latter authors also characterized the classes $L^{n,\infty}(\mathbb{R})$ and the class $AC^{n-1}(\mathbb{R})$ of functions with (n-1) absolutely continuous derivatives on $\mathbb{R}$ in a subsequent paper [15.8].

There are actually three problems formulated by Golomb in [15.6]. We state them as:

(I) Given $E \subset \mathbb{R}$ and a real-valued function f on E, under what conditions is there an $F \in L^{n,p}(\mathbb{R})$ such that $F(x) = f(x)$ for $x \in E$?

(II) To determine existence and uniqueness of an extremal $L^{n,p}(\mathbb{R})$-extension F_* for which $\int_{\mathbb{R}} |D^n F_*|^p$ is minimal.

(III) To characterize F_* as the solution of a multipoint boundary value problem.

In the case where E is a finite point set of cardinality at least n, problems I and II have affirmative answers as was established in [3.5] and the characterization of F_* is handled in even greater generality in section 3.3 of this monograph. However, the consideration of finite point sets is the key to a resolution of

(I) and (II). We have

Theorem 15.6. The function f with domain $E \subset \mathbb{R}$ has an $L^{n,p}(\mathbb{R})$ extension if and only if

$$(15.14) \qquad \sup_e \int_{\mathbb{R}} |D^m F_e|^p < \infty$$

where e denotes an arbitrary finite subset of E and

$$\|F_e^{(n)}\|_p = \inf\{\|F^{(n)}\|_p : F(x) = f(x),\ x \in e\}.$$

Moreover, if (15.14) holds then f has a unique extremal extension F_* which is the limit in $L^{n,p}(\mathbb{R})$ of any sequence F_{e_ν} where $e_\nu \subset e_{\nu+1}$, $\bigcup_\nu e_\nu$ is dense in E.

This theorem is an amalgam of Theorems 2.1 and 2.2 of [15.6]. The norm in the space $L^{n,p}(\mathbb{R})$ may conveniently be given by

$$\|f\| = \{\sum_{i=1}^{n} |f(x_i)|^p + \int_{\mathbb{R}} |f^{(n)}|^p\}^{1/p}$$

and implies uniform convergence of derivatives through order n - 1. The latter part of this theorem on the convergence of F_{e_ν} to F_* was proved in the case p = 2 by Golomb and Jerome [15.7] who studied questions (II) and (III) in this case. The notion of the essential boundary of E was also introduced by Golomb and Jerome as the set of limit points of E. In [15.6], this definition is extended, in the case a Taylor field is specified, to include points x of E for which $\mu(x) = n$. The essential boundary plays a role in uncoupling global characterizations described by (III). We sketch the ideas leading to this. Using essentially Theorem 3.1 of this monograph one proves

Theorem 15.7. A necessary and sufficient condition that the function $F \in L^{n,p}(\mathbb{R})$ is the extremal $L^{n,p}(\mathbb{R})$-interpolant of a Taylor field of variable height $\mu \leq n$ on the closed set E is

$$\int_{\mathbb{R}} |D^n F|^{p-1} \operatorname{sign}(D^n F) \cdot D^n G = 0$$

for every function $G \in L^{n,p}(\mathbb{R})$ for which

$$D^k G(x) = 0, \; k = 0,\dots,\mu(x) - 1, \; x \in E.$$

This is Theorem 4.1a of [15.6] and leads to the following solution of (III).

Theorem 15.8. If F is the extremal $L^{n,p}(\mathbb{R})$-interpolant of a Taylor field of height $\mu \leq n$ on the closed set E then

(i) $(D^n F)_s^{p-1} \in C^n(\mathbb{R}\backslash E)$; $D^k (D^n F)_s^{p-1}(x)$

exists and is continuous at $x \in \mathbb{R}\backslash E'$ for

(15.15) $\quad k = 0,\dots,n - 1 - \mu(x)$;

(ii) $D^n (D^n F)_s^{p-1}(x) = 0, \; x \in \mathbb{R}\backslash E$;

(iii) $D^n F(x) = 0$ for $x < \inf E$ and $x > \sup E$.

Here we have used the terminology

(15.16) $$(D^n F)_s^{p-1} = |D^n F|^{p-1} \operatorname{sign}(D^n F).$$

For p = 2, this result was obtained in [15.7]. There the idea was introduced of examining the sequence of multipoint problems obtained on the intervals of $\mathbb{R}\setminus E_b$, where E_b denotes the essential boundary of E. Golomb obtains the result (Theorem 5.1 of [15.6]) that the natural multipoint boundary value problem formulated from (15.15) on an interval J of $\mathbb{R}\setminus E_b$ has a unique solution which is the restriction of F to J. This completely solves (III).

A number of results are given for special sets E wherein the condition (15.14) is replaced by divided difference conditions similar to those of Theorem 15.5. For example, Golomb obtains the result

<u>Theorem 15.9</u>. Let $E = \{t_i\}_{i=-\infty}^{\infty}$ be a quasi-uniform partition satisfying

$$\delta \leq t_{i+1} - t_i \leq \frac{1}{\delta}, \ \delta > 0, \text{ for all } i.$$

Then f has an (extremal) extension F_* in $L^{n,p}(\mathbb{R})$ if and only if

$$\sum_{i=-\infty}^{\infty} |f(t_i,\ldots,t_{i+n})|^p < \infty. \tag{15.17}$$

An analogous result has been given by Smith [15.13] for $p = \infty$. The condition replacing (15.17) is the boundedness of the nth. order divided differences on E and hence is identical to the condition characterizing $L^{n,\infty}(\mathbb{R})$ functions described in [15.8].

<u>Remarks</u>. DeBoor shows in [15.1] that the sets N_i described in the Favard procedure have a natural form when the functions φ_i are (confluent) B-splines arising from interpolation at the points $a \leq t_1 \leq \ldots \leq t_{n+k} \leq b$ of a mesh with multiplicity not exceeding k; in this case, the N_i are unions of disjoint intervals $[t_L, t_R]$

where t_L and t_R belong to $\{t_i\}$ and, moreover, the following properties hold for the element $f \in W^{k,\infty}(a,b)$, interpolating the prescribed values at t_i, for which $f^{(k)}$ is the Favard solution:

(i) $f^{(k)}$ vanishes off (t_1, t_{n+k});

(ii) $|f^{(k)}|$ has all its jumps at $\{t_i\}_1^{n+k}$;

(iii) $f^{(k)}$ has less than n jumps in (t_1, t_{n+k}).

In [6.1], the authors defined an alternative piecewise perfect spline solution by taking the convex solution set of the original L^∞ minimization problem and systematically eliminating members by minimizing successively on (t_i, t_{i+1}). It was found that the process terminated with a unique piecewise perfect spline solution satisfying (ii) above, with no more than $k - 1$ jumps in $f^{(k)}$ on each (t_i, t_{i+1}).

The paper [15.7] of Golomb and Jerome was the first published investigation of problems II and III of section 15.3 and was done for $p = 2$. Since the analysis of [15.7] was carried out for general differential operators L, certain subtle questions arose in the boundary value problem characterization for F_*, which do not appear when $L = D^n$. The spaces considered in [15.7] owed much to earlier, unpublished, joint investigations of Golomb and Schoenberg.

In his dissertation, Smith [15.12] obtained a number of results similar to those described in section 15.3 for the Sobolev spaces $W^{n,p}(\mathbb{R})$. The reader will note that the spaces $L^{n,p}(\mathbb{R})$ possess some elements not in $L^p(\mathbb{R})$, so that the corresponding conditions for extremal extensions in $W^{n,p}(\mathbb{R})$ must insure that

$F_* \in L^p(\mathbb{R})$. For example, the analogue of Theorem 15.9 in $W^{n,p}(\mathbb{R})$ requires, in addition to (15.17), the condition that $\{f(t_i)\}_{i=-\infty}^{\infty} \in \ell^p$. Finally, we mention for completeness the papers of Coatmelec [15.3] and Glaeser [15.5] related to the Whitney extension problem. The latter paper, in fact, employs the device of perfect spline functions.

REFERENCES

15.1 C. DeBoor, "On best interpolation," J. of Approximation Theory, to appear.

15.2 C. K. Chui, P. W. Smith, and J. D. Ward, "Favard's solution is the limit of $W^{k,p}$-splines," manuscript.

15.3 C. Coatmelec, "Prolongement d'une fonction en une fonction differentiable. Diverses majorations sur le prolongement," in, Approximation with Special Emphasis on Spline Functions, I. J. Schoenberg, Ed., Academic Press, New York, 1969, pp. 29-49.

15.4 J. Favard, "Sur l'interpolation," J. Math. Pures et Appliques, 19 (1940), 281-306.

15.5 G. Glaeser, "Prolongement extremal de fonctions differentiables d'une variable," J. Approximation Theory, 8 (1973), 249-261.

15.6 M. Golomb, "$H^{m,p}$-extensions by $H^{m,p}$-splines," J. of Approximation Theory, 5 (1972), 238-275.

15.7 M. Golomb and J. Jerome, "Linear ordinary differential equations with boundary conditions on arbitrary point sets," Trans. Amer. Math. Soc., 153 (1971), 235-264.

15.8 J. W. Jerome and L. L. Schumaker, "Characterizations of absolute continuity and essential boundedness for higher order derivatives," J. Math. Anal. Appl., 42 (1973), 452-465.

15.9 A. N. Kolmogorov, "Über die beste Annäherung von Functionen einer gegebenen Funktionklasse," Ann. Math., (2), 37 (1936), 107-111.

15.10 F. Riesz, "Systeme integrierbarer Funktionen," Math. Ann., 69 (1910), 449-497.

15.11 I. J. Schoenberg, "Spline interpolation and the higher derivatives," Proc. Nat. Acad. Sci. U.S.A., 51 (1964), 24-28.

15.12 P. W. Smith, "$W^{r,p}(R)$-Splines," dissertation, Purdue University, Lafayette, Indiana, June, 1972.

15.13 ________, "$H^{r,\infty}(R)$ and $W^{r,\infty}(R)$-splines," Trans. Amer. Math. Soc., 192 (1974), 275-284.

15.14 V. M. Tihomirov, "On n-dimensional diameters of certain functional classes," Soviet Math.-Doklady, 1 (1960), 94-97.

15.15 ________, "Diameters of sets in function spaces and the theory of best approximations," Uspehi 15 No. 3 (93) (1960), 81-120.

15.16 ________, "Some problems of approximation theory," Soviet Math., 6 (1965), 202-205.

15.17 ________, "Best methods of approximation and interpolation of differentiable functions in the space C[-1,1]," Math. USSR - Sb., 9 (1969), 275-289.

15.18 H. Whitney, "Differentiable functions defined in closed sets." Trans. Amer. Math. Soc. 36 (1934), 369-337.

§16. Application of the Riesz-Fredholm-Schauder Theory to Spline Functions

We shall close the mongraph with an account of perhaps the tightest and most precise formulation of the spline theory achieveable to date. This is possible through the use of the Riesz-Fredholm-Schauder theory as elaborated by Aubin [10.1]. In Chapter Ten we had occasion to present, in Theorem 10.7, the fundamental underlying result of that theory and in this chapter we shall discuss, without proof, certain consequences. These results take the form of the equivalence of a generalized boundary value problem with a variational principle, connected by a generalized integration by parts formula, due to Aubin.

In his book, Aubin derived an abstract variational foundation for the finite element method. It was recognized by Jerome [3.4] that this theory totally characterizes generalized spline functions for non self-adjoint as well as self-adjoint bilinear forms. Subsequently, it was shown [16.6] that a theory could be developed, based totally on well-understood spectral properties, for the singular self-adjoint differential operators of mathematical physics. The purpose of this chapter is to indicate briefly how this can be done. We shall, however, first review the historical perspective out of which this arises.

In Chapters Two and Three (cf. especially section 3.3) we had occasion to discuss minimization problems whose solutions exhibit structural characteristics of so-called spline type. This subject has a rich history and it owes much of its development to the variety of equivalent extremal problems for which the spline, even the ordinary classical spline, arises as a solution. The reader can get a feeling for this by examining the excellent account in Chapter Three of Sard and Weintraub [16.10] (cf. also the review of the second author [Math. Comp. No. 121, 27 (1973)]). These developments, of course, predated the abstract Hilbert space formulation initiated, as mentioned in Chapter Two, by the French school following the lead of Golomb and Weinberger. In fairness, it should be recognized that the paper of DeBoor and Lynch [16.2], which exploited reproducing kernel Hilbert spaces, was a transition paper in this historical development. In 1967, Golomb [16.3] made the significant observation that the minimizing spline is in fact a distribution solution of the Euler equation of the quadratic functional $\int_a^b (Lf)^2$, L a nonsingular linear differential operator; this follows from the orthogonality relation $\int_a^b LsLu = 0$ satisfied by the interpolating spline solution s for all functions u vanishing at the nodal points. Thus, s is (locally) a distribution solution of $L^*Ls = 0$ and, by well understood ideas, is a classical solution. This latter fact, which we have made use of earlier in the monograph, is documented quite adequately in the book of Halperin [3.2]. Golomb's observation played an important part in the subsequent development of the Lg-splines [3.5] and he had clearly forseen such developments in his 1967 report. Golomb also developed in this paper, together with the second author, direct existence theorems on the minimization of $\|Lf\|_2$ over flats in $W^{n,2}(a,b)$, which then made possible the exploitation of the spline as a distribution. This observation of Golomb was a most significant event, now viewed retroactively, in the variational formulation of generalized spline functions. It is, for example, the point of view taken in section 10.2 and that which we now develop, first from an abstract point of view.

Let H and V be real Hilbert spaces and $B(\cdot,\cdot)$ a bilinear form on V such that

(i) $I : V \to H$ is continuous,

(16.1)

(ii) $B(\cdot,\cdot)$ is continuous on V, i.e.,

$$|B(u,v)| \leq C\|u\|_V\|v\|_V, \text{ for all } u, v \in V,$$

(iii) V is dense in H.

Let W be a Hilbert space and Γ a linear mapping of V into W such that

(16.2)

(i) Γ is a continuous mapping of V onto W,

(ii) the kernel V_0 of Γ is dense in H.

We have

Proposition 16.1. Suppose (16.1) and (16.2) are satisfied. Then there exists a closed linear operator Λ in H with domain $D_\Lambda \subset V$ dense in V and H satisfying

(16.3) $$B(u,v) = (\Lambda u,v)_H \text{ for all } u \in D_\Lambda,\ v \in V_0.$$

Moreover, there exists a continuous, uniquely determined linear operator Ω mapping D_Λ into W such that

(16.4) $$B(u,v) = (\Lambda u,v)_H + (\Omega u,\Gamma v)_W, \text{ for all } u \in D_\Lambda,\ v \in V.$$

This proposition is proved in [10.1, Theorem 2-1, p. 174] and also in [3.4, Lemma 4.1]. The generalized integration by parts formula (16.4) plays a key role in the sequel.

Proposition 16.2. Let P be an orthogonal projection of W into itself, let Q = I - P and let U_0 denote the kernel of $P\Gamma$. If (16.1) and (16.2) hold then the boundary value problem

$$\text{(16.5)}\quad \begin{aligned} &\text{(i)} \quad \Lambda s = 0 \\ &\text{(ii)} \quad P\Gamma s = u \\ &\text{(iii)} \quad Q\Omega s = v \end{aligned}$$

has the solution $s \in D_\Lambda$, where $u \in PW$ and $v \in QW$, if and only if s satisfies the projection relation

$$\text{(16.6)} \qquad B(s,f) = (v, Q\Gamma f)_W, \text{ for all } f \in U_0.$$

This proposition is proved in [10.1, Theorem 2-2, p. 178] and in [3.4, Lemma 4.2]. It does not assert the existence of solutions of (16.5); rather, it asserts the simultaneous existence of solutions of (16.5) and (16.6). In order to obtain solutions of (16.5), we require additional hypotheses.

(16.7)

(i) $I : V \to H$ is compact,

(ii) there exist positive constants C and C_1 such that

$$B(u,u) + C(u,u)_H \geq C_1(u,u)_V, \text{ for all } u \in V.$$

We have

<u>Proposition 16.3</u>. Suppose that (16.1), (16.2) and (16.7) are satisfied. Then, if the only solution $f \in D_\Lambda$ to the boundary value problem

$$\text{(i)} \quad \Lambda f = 0$$

$$(16.8) \quad \text{(ii)} \quad P\Gamma f = 0$$

$$\text{(iii)} \quad Q\Omega f = 0$$

is the zero solution, (16.5) has a solution $s \in D_\Lambda$ for every $u \in PW$ and $v \in QW$. If (16.8) has nontrivial solutions, then (16.5) has a solution if and only if u and v satisfy the compatibility condition

$$(v, Q\Gamma f)_W = (u, P\omega f)_W$$

for all nontrivial solutions f of the adjoint problem

$$\text{(i)} \quad \overline{\Lambda} f = 0$$

$$(16.9) \quad \text{(ii)} \quad P\Gamma f = 0$$

$$\text{(iii)} \quad Q\omega f = 0$$

Here $\overline{\Lambda}$ and ω arise from (16.4) when $B(\cdot,\cdot)$ is replaced by $\overline{B}(\cdot,\cdot)$, where $\overline{B}(g,h) = B(h,g)$ on V.

This proposition is proved in [10.1, Theorem 2-3, p. 181] and [3.4, Lemma 4.3]. Essential use is made, in the proof, of Theorem 10.7.

The implementation of these results requires of course the choice of spaces and bilinear forms for which the hypotheses (16.1), (16.2) and (16.7) hold. Thus, if b_{nn} is a positive function on $I = (a,b)$ satisfying $1/b_{nn} \in L^1(I)$, we may define V by

$$(16.10) \quad V = \{u \in C^{n-1}(\bar{I}) : D^{n-1}u \text{ is absolutely continuous and } \sqrt{b_{nn}}\, D^n u \in L^2(I)\}.$$

A bilinear form may be defined on V in a natural way. Thus, for $(i,j) \in \{0,1,\ldots,n\} \times \{0,1,\ldots,n\} - (n,n)$, let b_{ij} be a real-valued function satisfying

$$(16.11) \quad \begin{aligned} & b_{ij} \in L^\infty(I), \ i \text{ and } j < n \\ & b_{ij}/\sqrt{b_{nn}} \in L^\infty(I) \text{ if } i = n \text{ or } j = n. \end{aligned}$$

We define,

$$B(u,v) = \sum_{i,j=0}^{n} \int_I b_{ij} D^i u D^j v, \quad u, v \in V.$$

With the choice $H = L^2(I)$, it is shown in [16.7] that (16.1) and (16.7) hold through the use of generalized Sobolev and interpolation inequalities, first proved in [16.4]. Moreover, it is shown in [16.5] that the Extended-Hermite-Birkhoff linear functionals with nodal points in $\bar{I}$ are continuous on V and it follows easily that (16.2) holds with $W = \mathbb{R}^k$, k the cardinality of the set of E-H-B functionals. In [16.5], the symmetric bilinear functional for which $b_{ij} \equiv 0$, $i \neq j$, in (16.11) is utilized and in [3.4] the choice $b_{nn} \equiv 1$ is made. In both papers, characterizations are presented, which we shall not repeat here, for the problem (16.5) in the case where $P\Gamma$ is represented as a Cartesian product of the given E-H-B functionals on the space V, defined by (16.10). In this case the operator $Q\Omega$ is the Cartesian product of operators $[R_{ij}(\cdot)]$ as in section 3.3. In [16.5], v = 0 is chosen in (16.5iii). In this case, it is possible to express a minimization principle for solutions of (16.5) provided $B(\cdot,\cdot)$ is nonnegative and symmetric. This result was obtained in [16.6] and we express it formally as

Theorem 16.4. Suppose (16.1), (16.2) and (16.7) hold and that $B(\cdot,\cdot)$ is symmetric and nonnegative (definite) on V. Then a solution s of (16.5) always exists for the case v = 0 and moreover

$$B(s,s) = \inf\{B(f,f) : f \in V,\ P\Gamma f = u\}.$$

The approach outlined above explicitly defines V, H and W and proceeds under the assumption that (16.1), (16.2) and (16.7) hold for a given bilinear form $B(\cdot,\cdot)$ and observation operator Γ. However, quite often an explicit definition of V is not immediately available. This is the case for the singular differential operators of mathematical physics, which are defined on dense linear subspaces of $L^2(J)$ for a bounded or unbounded interval J in $\mathbb{R}$. The only information readily available in these cases is spectral information; for example, it is very often known that Λ has a compact resolvent. It was the purpose of [16.6] to devise such a theory for symmetric bilinear forms. As a preliminary result we have

Proposition 16.5. Let H be a real Hilbert space, let V be a dense linear subspace of H and let B be a symmetric bilinear form on V satisfying (16.7ii). Suppose V is a Hilbert space under the inner product

$$(16.12) \qquad [f,g] = B(f,g) + C(f,g)_H,$$

where C is given by (16.7ii). Then, if the Friedrich's self-adjoint operator A defined by [16.9, p. 335]

$$(16.13) \qquad [f,g] = (Af,g)_H \text{ for all } f \in D_A,\ g \in V$$

has a compact resolvent, it follows that (16.7i) holds and thus (16.1) and (16.7) hold.

Suppose now that M is a given formally self-adjoint differential operator with real coefficients of the form

$$(16.14) \qquad M = \sum_{j=0}^{n} (-1)^j D^j (a_j D^j)$$

where $a_j \in C^j(J)$ and $a_n(x) \neq 0$ if $x \in J$. Here J is an arbitrary subinterval of $\mathbb{R}$, not necessarily bounded or even compact. Let $D_M = \{f \in AC^{2n-1}(J) : f, Mf \in L^2(J)\}$. We have the following result.

Theorem 16.6. Suppose that there is a self-adjoint restriction A of M in $L^2(J)$ with a compact resolvent such that the spectrum of A is bounded from below. Suppose in addition that the coefficients of M satisfy

$$(16.15) \qquad \begin{aligned} &a_n(x) > 0,\ x \in J, \\ &a_j(x) \geq 0,\ x \in J,\ 0 \leq j \leq n-1. \end{aligned}$$

Then the linear space V,

$$(16.16)\quad \begin{aligned} &\text{(i)}\quad V = \{f \in AC^{n-1}(J) : B(f,f) < \infty\}, \\ &\text{(ii)}\quad B(f,g) = \sum_{j=0}^{m} \int_J a_j D^j f D^j g, \end{aligned}$$

is a Hilbert space and (16.1) and (16.7) hold with $H = L^2(J)$. Furthermore, if K is any compact subinterval of J then the mapping

$$V \to W^{2,n}(K) \; : \; f \to f|_K$$

is continuous; hence any continuous linear functional θ on the Sobolev space $W^{2,n}(K)$ can be extended to a continuous linear functional $\tilde{\theta}$ on V by

$$\tilde{\theta} f = \theta(f|_K).$$

Proposition 16.5 and Theorem 16.6 are proved in [16.6] where an application is presented with (E-H-B) linear functionals at interior points of J.

Remarks. The 1966 Grenoble dissertation of Atteia as well as a preprint of the paper of Anselone and Laurent [2.1] were available to Golomb during the summer of 1967 when the report [16.3] was prepared. Golomb's approach was likely motivated by the attempt to distinguish clearly between questions of existence and uniqueness; these authors and earlier ones had assumed uniqueness a priori and established existence on the basis of uniqueness. However, as the earlier discussion has suggested, the idea of the spline as a distribution has led to (local) classical differential equation characterizations for the solution s of projection relations of the form (16.6), at least on intervals not contained in the support of

PΓ.

In 1967 Golomb had still another fruitful idea. The paper [15.7] was being planned at that time to consider questions (II) and (III) of section 15.3 for the extremal solutions of $\int_{\mathbb{R}} (Lf)^2$ and Golomb suggested consideration of more general forms $\sum_q \int_{\mathbb{R}} (L_q f)^2$ for a sequence of differential operators $L_1,\ldots,L_k$ for which L_k is uniquely of maximal order. This in fact was implemented in [15.7]. We should mention also the interesting paper of Lucas [16.8], which appeared simultaneously with [16.5], in which general bilinear forms were studied. The role of the generalized spline in the solution of multipoint boundary value problems, particularly in the variational definition of the Green's function, was carried out in [16.4], although the underlying ideas were latent in [15.7].

The reader familiar with the Fredholm alternative may wonder why the alternative has been expressed in terms of the adjoint problem (16.9) rather than an operator adjoint. In fact, these coincide; the adjoint of the restriction of Λ to the linear subspace of D_Λ satisfying (16.8ii) and (16.8iii) is precisely the restriction of $\overline{\Lambda}$ to the subspace satisfying (16.9ii) and (16.9iii), provided this restriction of Λ is densely defined. The hypothesis that the restriction $\hat{\Lambda}$ of Λ is densely defined is therefore assumed in this chapter. It is satisfied in our application with E-H-B functionals and whenever $B(\cdot,\cdot)$ is symmetric and nonnegative on a separable Hilbert space. Since $\hat{\Lambda}$ is also closed in H, its dense domain of definition implies [16.9, p. 305] that its adjoint is densely defined and that $\hat{\Lambda}^{**} = \hat{\Lambda}$.

Finally, we wish to emphasize that this chapter and, more generally, the monograph as a whole has not attempted an historical account of the subject of variational splines. We leave this to books already published and in preparation. However, some perspective can be gained from the book of Ahlberg, Nilson and Walsh [16.1], the content and references of the paper of Schultz and Varga [10.6] and the brief summary in [2.7].

REFERENCES

16.1 J. H. Ahlberg, E. N. Nilson and J. L. Walsh, The Theory of Splines and Their Applications, Academic Press, New York, 1967.

16.2 C. DeBoor and R. E. Lynch, "On splines and their minimum properties," J. Math. Mech., 15 (1966), 953-969.

16.3 M. Golomb, "Splines, n-widths, and optimal approximations," MRC Tech. Summ. Rpt. No. 784, Mathematics Research Center, Madison, Wisconsin, September, 1967.

16.4 J. W. Jerome, "Singular, self-adjoint, multipoint boundary value problems: solutions and approximations," in, Linear Operators and Approximation (P. L. Butzer, J. P. Kahane and B. Z.-Nagy, editors), Birkhäuser-Verlag, Basel (1972), 470-486.

16.5 J. Jerome and J. Pierce, "On spline functions determined by singular self-adjoint differential operators," J. Approximation Theory, 5 (1972), 15-40.

16.6 J. W. Jerome, "On spline functions derivable from singular differential operators with compact resolvents," J. Math. Anal. Appl., to appear.

16.7 ________, "Generalized boundary value problems and the evolution equation," manuscript.

16.8 T. R. Lucas, "M-splines," J. Approximation Theory, 5 (1972), 1-14.

16.9 F. Riesz and B. Sz.-Nagy, Functional Analysis, Ungar, New York, 1955.

16.10 A. Sard and S. Weintraub, A Book of Splines, Wiley, New York, 1971.

§17. Epilogue

The thrust of this monograph has been along the branches of the curvature problems in L^2 and L^∞ on the one hand and that of bang-bang properties of solutions of various L^∞ extremal problems on the other hand. Berkovitz and Pollard, in a series of papers [17.1, 17.2, 17.3, 17.4], have shown that the bang-bang splines arise quite naturally in the analysis of filtering problems. We have commented earlier on the well-understood connections with control problems. One of the rich sources for investigation is the highly nonlinear problem of the determination of the switching points of the bang-bang controls.

The challenging problem of the efficient computation of the elastica remains. There have been numerous investigations already into the problem of the calculation of stable equilibrium configurations. These are summarized in the informative report of Michael Malcolm [17.7]. In addition to the references of Chapter Nine we may cite the work of Glass [17.5], Larkin [17.6], Mehlum [17.8] and Woodford [17.9] as well as Malcolm's work. None of these authors has considered, however, the general problem with Lagrange multipliers, necessitated because of the lack of stable equilibrium configurations for certain data arrays. This seems a challenging problem for the future.

REFERENCES

17.1 L. Berkovitz and H. Pollard, "A non-classical variational problem arising from an optimal filter problem," _Arch. Rational Mech. and Analysis_, 26 (1967), 281-304.

17.2 __________, "A non-classical variational problem arising from an optimal filter problem II," _Arch. Rational Mech. and Analysis_, 38 (1970), 161-172.

17.3 __________, "A variational problem related to an optimal filter problem with self-correlated noise," _Trans. Amer. Math. Soc._, 142 (1969), 153-175.

17.4 ________, "Addenda to [17.3]," Trans. Amer. Math. Soc., 157 (1971), 499-504.

17.5 J. M. Glass, "Smooth-curve interpolation: A generalized spline fit procedure," BIT, 6 (1966), 277-293.

17.6 F. M. Larkin, "An interpolation procedure based on fitting elasticas to given data points," Culham Operating System, Note 5/66, Theory Division, Culham Laboratory, Berkshire, England, 1966.

17.7 M. A. Malcolm, "Nonlinear spline functions," Report 73-372, Computer Science Department, Stanford University, June, 1973.

17.8 E. Mehlum, "Curve and surface fitting based on variational criteria for smoothness," dissertation, University of Oslo, 1969.

17.9 C. H. Woodford, "Smooth curve interpolation," BIT, 9 (1969), 69-77.

Subject Index

Vol. 309: D. H. Sattinger, Topics in Stability and Bifurcation Theory. VI, 190 pages. 1973. DM 20,–

Vol. 310: B. Iversen, Generic Local Structure of the Morphisms in Commutative Algebra. IV, 108 pages. 1973. DM 18,–

Vol. 311: Conference on Commutative Algebra. Edited by J. W. Brewer and E. A. Rutter. VII, 251 pages. 1973. DM 24,–

Vol. 312: Symposium on Ordinary Differential Equations. Edited by W. A. Harris, Jr. and Y. Sibuya. VIII, 204 pages. 1973. DM 22,–

Vol. 313: K. Jörgens and J. Weidmann, Spectral Properties of Hamiltonian Operators. III, 140 pages. 1973. DM 18,–

Vol. 314: M. Deuring, Lectures on the Theory of Algebraic Functions of One Variable. VI, 151 pages. 1973. DM 18,–

Vol. 315: K. Bichteler, Integration Theory (with Special Attention to Vector Measures). VI, 357 pages. 1973. DM 29,–

Vol. 316: Symposium on Non-Well-Posed Problems and Logarithmic Convexity. Edited by R. J. Knops. V, 176 pages. 1973. DM 20,–

Vol. 317: Séminaire Bourbaki – vol. 1971/72. Exposés 400–417. IV, 361 pages. 1973. DM 29,–

Vol. 318: Recent Advances in Topological Dynamics. Edited by A. Beck. VIII, 285 pages. 1973. DM 27,–

Vol. 319: Conference on Group Theory. Edited by R. W. Gatterdam and K. W. Weston. V, 188 pages. 1973. DM 20,–

Vol. 320: Modular Functions of One Variable I. Edited by W. Kuyk. V, 195 pages. 1973. DM 20,–

Vol. 321: Séminaire de Probabilités VII. Edité par P. A. Meyer. VI, 322 pages. 1973. DM 29,–

Vol. 322: Nonlinear Problems in the Physical Sciences and Biology. Edited by I. Stakgold, D. D. Joseph and D. H. Sattinger. VIII, 357 pages. 1973. DM 29,–

Vol. 323: J. L. Lions, Perturbations Singulières dans les Problèmes aux Limites et en Contrôle Optimal. XII, 645 pages. 1973. DM 46,–

Vol. 324: K. Kreith, Oscillation Theory. VI, 109 pages. 1973. DM 18,–

Vol. 325: C.-C. Chou, La Transformation de Fourier Complexe et L'Equation de Convolution. IX, 137 pages. 1973. DM 18,–

Vol. 326: A. Robert, Elliptic Curves. VIII, 264 pages. 1973. DM 24,–

Vol. 327: E. Matlis, One-Dimensional Cohen-Macaulay Rings. XII, 157 pages. 1973. DM 20,–

Vol. 328: J. R. Büchi and D. Siefkes, The Monadic Second Order Theory of All Countable Ordinals. VI, 217 pages. 1973. DM 22,–

Vol. 329: W. Trebels, Multipliers for (C, α)-Bounded Fourier Expansions in Banach Spaces and Approximation Theory. VII, 103 pages. 1973. DM 18,–

Vol. 330: Proceedings of the Second Japan-USSR Symposium on Probability Theory. Edited by G. Maruyama and Yu. V. Prokhorov. VI, 550 pages. 1973. DM 40,–

Vol. 331: Summer School on Topological Vector Spaces. Edited by L. Waelbroeck. VI, 226 pages. 1973. DM 22,–

Vol. 332: Séminaire Pierre Lelong (Analyse) Année 1971-1972. V, 131 pages. 1973. DM 18,–

Vol. 333: Numerische, insbesondere approximationstheoretische Behandlung von Funktionalgleichungen. Herausgegeben von R. Ansorge und W. Törnig. VI, 296 Seiten. 1973. DM 27,–

Vol. 334: F. Schweiger, The Metrical Theory of Jacobi-Perron Algorithm. V, 111 pages. 1973. DM 18,–

Vol. 335: H. Huck, R. Roitzsch, U. Simon, W. Vortisch, R. Walden, B. Wegner und W. Wendland, Beweismethoden der Differentialgeometrie im Großen. IX, 159 Seiten. 1973. DM 20,–

Vol. 336: L'Analyse Harmonique dans le Domaine Complexe. Edité par E. J. Akutowicz. VIII, 169 pages. 1973. DM 20,–

Vol. 337: Cambridge Summer School in Mathematical Logic. Edited by A. R. D. Mathias and H. Rogers. IX, 660 pages. 1973. DM 46,–

Vol: 338: J. Lindenstrauss and L. Tzafriri, Classical Banach Spaces. IX, 243 pages. 1973. DM 24,–

Vol. 339: G. Kempf, F. Knudsen, D. Mumford and B. Saint-Donat, Toroidal Embeddings I. VIII, 209 pages. 1973. DM 22,–

Vol. 340: Groupes de Monodromie en Géométrie Algébrique. (SGA 7 II). Par P. Deligne et N. Katz. X, 438 pages. 1973. DM 44,–

Vol. 341: Algebraic K-Theory I, Higher K-Theories. Edited by H. Bass. XV, 335 pages. 1973. DM 29,–

Vol. 342: Algebraic K-Theory II, "Classical" Algebraic K-Theory, and Connections with Arithmetic. Edited by H. Bass. XV, 527 pages. 1973. DM 40,–

Vol. 343: Algebraic K-Theory III, Hermitian K-Theory and Geometric Applications. Edited by H. Bass. XV, 572 pages. 1973. DM 40,–

Vol. 344: A. S. Troelstra (Editor), Metamathematical Investigation of Intuitionistic Arithmetic and Analysis. XVII, 485 pages. 1973. DM 38,–

Vol. 345: Proceedings of a Conference on Operator Theory. Edited by P. A. Fillmore. VI, 228 pages. 1973. DM 22,–

Vol. 346: Fučík et al., Spectral Analysis of Nonlinear Operators. II, 287 pages. 1973. DM 26,–

Vol. 347: J. M. Boardman and R. M. Vogt, Homotopy Invariant Algebraic Structures on Topological Spaces. X, 257 pages. 1973. DM 24,–

Vol. 348: A. M. Mathai and R. K. Saxena, Generalized Hypergeometric Functions with Applications in Statistics and Physical Sciences. VII, 314 pages. 1973. DM 26,–

Vol. 349: Modular Functions of One Variable II. Edited by W. Kuyk and P. Deligne. V, 598 pages. 1973. DM 38,–

Vol. 350: Modular Functions of One Variable III. Edited by W. Kuyk and J.-P. Serre. V, 350 pages. 1973. DM 26,–

Vol. 351: H. Tachikawa, Quasi-Frobenius Rings and Generalizations. XI, 172 pages. 1973. DM 20,–

Vol. 352: J. D. Fay, Theta Functions on Riemann Surfaces. V, 137 pages. 1973. DM 18,–

Voi. 353: Proceedings of the Conference on Orders, Group Rings and Related Topics. Organized by J. S. Hsia, M. L. Madan and T. G. Ralley. X, 224 pages. 1973. DM 22,–

Vol. 354: K. J. Devlin, Aspects of Constructibility. XII, 240 pages. 1973. DM 24,–

Vol. 355: M. Sion, A Theory of Semigroup Valued Measures. V, 140 pages. 1973. DM 18,–

Vol. 356: W. L. J. van der Kallen, Infinitesimally Central-Extensions of Chevalley Groups. VII, 147 pages. 1973. DM 18,–

Vol. 357: W. Borho, P. Gabriel und R. Rentschler, Primideale in Einhüllenden auflösbarer Lie-Algebren. V, 182 Seiten. 1973. DM 20,–

Vol. 358: F. L. Williams, Tensor Products of Principal Series Representations. VI, 132 pages. 1973. DM 18,–

Vol. 359: U. Stammbach, Homology in Group Theory. VIII, 183 pages. 1973. DM 20,–

Vol. 360: W. J. Padgett and R. L. Taylor, Laws of Large Numbers for Normed Linear Spaces and Certain Fréchet Spaces. VI, 111 pages. 1973. DM 18,–

Vol. 361: J. W. Schutz, Foundations of Special Relativity: Kinematic Axioms for Minkowski Space Time. XX, 314 pages. 1973. DM 26,–

Vol. 362: Proceedings of the Conference on Numerical Solution of Ordinary Differential Equations. Edited by D. Bettis. VIII, 490 pages. 1974. DM 34,–

Vol. 363: Conference on the Numerical Solution of Differential Equations. Edited by G. A. Watson. IX, 221 pages. 1974. DM 20,–

Vol. 364: Proceedings on Infinite Dimensional Holomorphy. Edited by T. L. Hayden and T. J. Suffridge. VII, 212 pages. 1974. DM 20,–

Vol. 365: R. P. Gilbert, Constructive Methods for Elliptic Equations. VII, 397 pages. 1974. DM 26,–

Vol. 366: R. Steinberg, Conjugacy Classes in Algebraic Groups (Notes by V. V. Deodhar). VI, 159 pages. 1974. DM 18,–

Vol. 367: K. Langmann und W. Lütkebohmert, Cousinverteilungen und Fortsetzungssätze. VI, 151 Seiten. 1974. DM 16,–

Vol. 368: R. J. Milgram, Unstable Homotopy from the Stable Point of View. V, 109 pages. 1974. DM 16,–

Vol. 369: Victoria Symposium on Nonstandard Analysis. Edited by A. Hurd and P. Loeb. XVIII, 339 pages. 1974. DM 26,–

Vol. 370: B. Mazur and W. Messing, Universal Extensions and One Dimensional Crystalline Cohomology. VII, 134 pages. 1974. DM 16,-

Vol. 371: V. Poenaru, Analyse Différentielle. V, 228 pages. 1974. DM 20,-

Vol. 372: Proceedings of the Second International Conference on the Theory of Groups 1973. Edited by M. F. Newman. VII, 740 pages. 1974. DM 48,-

Vol. 373: A. E. R. Woodcock and T. Poston, A Geometrical Study of the Elementary Catastrophes. V, 257 pages. 1974. DM 22,-

Vol. 374: S. Yamamuro, Differential Calculus in Topological Linear Spaces. IV, 179 pages. 1974. DM 18,-

Vol. 375: Topology Conference 1973. Edited by R. F. Dickman Jr. and P. Fletcher. X, 283 pages. 1974. DM 24,-

Vol. 376: D. B. Osteyee and I. J. Good, Information, Weight of Evidence, the Singularity between Probability Measures and Signal Detection. XI, 156 pages. 1974. DM 16.-

Vol. 377: A. M. Fink, Almost Periodic Differential Equations. VIII, 336 pages. 1974. DM 26,-

Vol. 378: TOPO 72 - General Topology and its Applications. Proceedings 1972. Edited by R. Alò, R. W. Heath and J. Nagata. XIV, 651 pages. 1974. DM 50,-

Vol. 379: A. Badrikian et S. Chevet, Mesures Cylindriques, Espaces de Wiener et Fonctions Aléatoires Gaussiennes. X, 383 pages. 1974. DM 32,-

Vol. 380: M. Petrich, Rings and Semigroups. VIII, 182 pages. 1974. DM 18,-

Vol. 381: Séminaire de Probabilités VIII. Edité par P. A. Meyer. IX, 354 pages. 1974. DM 32,-

Vol. 382: J. H. van Lint, Combinatorial Theory Seminar Eindhoven University of Technology. VI, 131 pages. 1974. DM 18,-

Vol. 383: Séminaire Bourbaki - vol. 1972/73. Exposés 418-435 IV, 334 pages. 1974. DM 30,-

Vol. 384: Functional Analysis and Applications, Proceedings 1972. Edited by L. Nachbin. V, 270 pages. 1974. DM 22,-

Vol. 385: J. Douglas Jr. and T. Dupont, Collocation Methods for Parabolic Equations in a Single Space Variable (Based on C^1-Piecewise-Polynomial Spaces). V, 147 pages. 1974. DM 16,-

Vol. 386: J. Tits, Buildings of Spherical Type and Finite BN-Pairs. IX, 299 pages. 1974. DM 24,-

Vol. 387: C. P. Bruter, Eléments de la Théorie des Matroïdes. V, 138 pages. 1974. DM 18,-

Vol. 388: R. L. Lipsman, Group Representations. X, 166 pages. 1974. DM 20,-

Vol. 389: M.-A. Knus et M. Ojanguren, Théorie de la Descente et Algèbres d' Azumaya. IV, 163 pages. 1974. DM 20,-

Vol. 390: P. A. Meyer, P. Priouret et F. Spitzer, Ecole d'Eté de Probabilités de Saint-Flour III - 1973. Edité par A. Badrikian et P.-L. Hennequin. VIII, 189 pages. 1974. DM 20,-

Vol. 391: J. Gray, Formal Category Theory: Adjointness for 2-Categories. XII, 282 pages. 1974. DM 24,-

Vol. 392: Géométrie Différentielle, Colloque, Santiago de Compostela, Espagne 1972. Edité par E. Vidal. VI, 225 pages. 1974. DM 20,-

Vol. 393: G. Wassermann, Stability of Unfoldings. IX, 164 pages. 1974. DM 20,-

Vol. 394: W. M. Patterson 3rd, Iterative Methods for the Solution of a Linear Operator Equation in Hilbert Space - A Survey. III, 183 pages. 1974. DM 20,-

Vol. 395: Numerische Behandlung nichtlinearer Integrodifferential- und Differentialgleichungen. Tagung 1973. Herausgegeben von R. Ansorge und W. Törnig. VII, 313 Seiten. 1974. DM 28,-

Vol. 396: K. H. Hofmann, M. Mislove and A. Stralka, The Pontryagin Duality of Compact O-Dimensional Semilattices and its Applications. XVI, 122 pages. 1974. DM 18,-

Vol. 397: T. Yamada, The Schur Subgroup of the Brauer Group. V, 159 pages. 1974. DM 18,-

Vol. 398: Théories de l'Information, Actes des Rencontres de Marseille-Luminy, 1973. Edité par J. Kampé de Fériet et C. Picard. XII, 201 pages. 1974. DM 23,-

Vol. 399: Functional Analysis and its Applications, Proceedings 1973. Edited by H. G. Garnir, K. R. Unni and J. H. Williamson. XVII, 569 pages. 1974. DM 44,-

Vol. 400: A Crash Course on Kleinian Groups - San Francisco 1974. Edited by L. Bers and I. Kra. VII, 130 pages. 1974. DM 18,-

Vol. 401: F. Atiyah, Elliptic Operators and Compact Groups. V, 93 pages. 1974. DM 18,-

Vol. 402: M. Waldschmidt, Nombres Transcendants. VIII, 277 pages. 1974. DM 25,-

Vol. 403: Combinatorial Mathematics - Proceedings 1972. Edited by D. A. Holton. VIII, 148 pages. 1974. DM 18,-

Vol. 404: Théorie du Potentiel et Analyse Harmonique. Edité par J. Faraut. V, 245 pages. 1974. DM 25,-

Vol. 405: K. Devlin and H. Johnsbråten, The Souslin Problem. VIII, 132 pages. 1974. DM 18,-

Vol. 406: Graphs and Combinatorics - Proceedings 1973. Edited by R. A. Bari and F. Harary. VIII, 355 pages. 1974. DM 30,-

Vol. 407: P. Berthelot, Cohomologie Cristalline des Schémas de Caracteristique $p > o$. VIII, 598 pages. 1974. DM 44,-

Vol. 408: J. Wermer, Potential Theory. VIII, 146 pages. 1974. DM 18,-

Vol. 409: Fonctions de Plusieurs Variables Complexes, Séminaire François Norguet 1970-1973. XIII, 612 pages. 1974. DM 47,-

Vol. 410: Séminaire Pierre Lelong (Analyse) Année 1972-1973. VI, 181 pages. 1974. DM 18,-

Vol. 411: Hypergraph Seminar. Ohio State University, 1972. Edited by C. Berge and D. Ray-Chaudhuri. IX, 287 pages. 1974. DM 28,-

Vol. 412: Classification of Algebraic Varieties and Compact Complex Manifolds. Proceedings 1974. Edited by H. Popp. V, 333 pages. 1974. DM 30,-

Vol. 413: M. Bruneau, Variation Totale d'une Fonction. XIV, 332 pages. 1974. DM 30,-

Vol. 414: T. Kambayashi, M. Miyanishi and M. Takeuchi, Unipotent Algebraic Groups. VI, 165 pages. 1974. DM 20,-

Vol. 415: Ordinary and Partial Differential Equations, Proceedings of the Conference held at Dundee, 1974. XVII, 447 pages. 1974. DM 37,-

Vol. 416: M. E. Taylor, Pseudo Differential Operators. IV, 155 pages. 1974. DM 18,-

Vol. 417: H. H. Keller, Differential Calculus in Locally Convex Spaces. XVI, 131 pages. 1974. DM 18,-

Vol. 418: Localization in Group Theory and Homotopy Theory and Related Topics Battelle Seattle 1974 Seminar. Edited by P. J. Hilton. VI, 171 pages. 1974. DM 20,-

Vol. 419: Topics in Analysis - Proceedings 1970. Edited by O. E. Lehto, I. S. Louhivaara, and R. H. Nevanlinna. XIII, 391 pages. 1974. DM 35,-

Vol. 420: Category Seminar. Proceedings, Sydney Category Theory Seminar 1972/73. Edited by G. M. Kelly. VI, 375 pages. 1974. DM 32,-

Vol. 421: V. Poénaru, Groupes Discrets. VI, 216 pages. 1974. DM 23,-

Vol. 422: J.-M. Lemaire, Algèbres Connexes et Homologie des Espaces de Lacets. XIV, 133 pages. 1974. DM 23,-

Vol. 423: S. S. Abhyankar and A. M. Sathaye, Geometric Theory of Algebraic Space Curves. XIV, 302 pages. 1974. DM 28,-

Vol. 424: L. Weiss and J. Wolfowitz, Maximum Probability Estimators and Related Topics. V, 106 pages. 1974. DM 18,-

Vol. 425: P. R. Chernoff and J. E. Marsden, Properties of Infinite Dimensional Hamiltonian Systems. IV, 160 pages. 1974. DM 20,-

Vol. 426: M. L. Silverstein, Symmetric Markov Processes. IX, 287 pages. 1974. DM 28,-

Vol. 427: H. Omori, Infinite Dimensional Lie Transformation Groups. XII, 149 pages. 1974. DM 18,-

Vol. 428: Algebraic and Geometrical Methods in Topology, Proceedings 1973. Edited by L. F. McAuley. XI, 280 pages. 1974. DM 28,-